HELIOSTAT FIELDS

HELIOSTAT FIELDS
A Classical Approach to Analysis and Layout Optimization

Shanley Lutchman

ALTARIAH BOOKS

AN ALTARIAH BOOKS LTD., PUBLICATION

Published by Altariah Books, Ltd., Pretoria, South Africa.

Heliostat Fields / Shanley Lutchman
ISBN 978-0-620-71834-9

CONTENTS

LIST OF FIGURES

LIST OF TABLES

NOMENCLATURE

English Letters

A	surface area
B	day angle
D	diameter
d	distance
d_c	critical distance
E	equation of time
H	height
H_h	heliostat height
H_w	heliostat width
I	intercepted energy
k	hour
L	length
L_{loc}	longitude at location
L_{st}	longitude of local time zone
$\mathbf{N}$	heliostat normal

$\mathbf{n}$	unitized heliostat normal
m	number of heliostats
n	number of design variables
Q	energy
$\mathbf{S}$	sun vector
$\mathbf{s}$	unitized sun vector
$\mathbf{T}$	target vector
$\mathbf{t}$	unitized target vector
t	scalar parameter of 3-dimensional object equations
x	set of function input variables

Greek Letters

α_s	solar altitude angle
γ_s	solar azimuth angle
δ	declination angle
δ_h	discretization node separation distance along height
δ_h	discretization node separation distance along width
η_a	atmospheric attenuation efficiency
η_b	blocking efficiency
η_c	cosine efficiency
η_s	shading efficiency
η_{sp}	spillage efficiency
ϕ	latitude angle
θ_z	zenith angle
ω	hour angle

Subscripts

E	East
i	heliostat number
z	zenith
h	hour
N	North
T	target

Constants

π	3.14

λ 9.3 mrad

Abbreviations

CSP   concentrating solar power
DNI   direct normal irradiation
HPC   high performance computing cluster
SAO   sequential approximate optimization

INTRODUCTION

Renewable energy technology makes it possible to generate power in a manner that has significantly lower adverse environmental impacts than conventional energy technology. Heliostat field layout optimization for central receivers forms part of the research that is being conducted in the area of renewable energy technology. The following sections describe how the information in this book fits into renewable energy technology research, what the objective of the approach presented in this book is and what has been done to meet this objective.

Background

There are various renewable energy resources available for power generation including solar, wind, hydro and geothermal energy. The contents of this book relates to technology used in the conversion of solar energy to electrical energy.

Central Receiver Systems

Concentrating solar power (CSP) is a method of harnessing the energy from the sun, which uses mirrors that are placed in such a way that they can reflect light from the sun to a centralized area. The sun's rays, instead of being absorbed by the earth, are reflected to that particular area. The central receiver system is an application of

Figure I.1: The Gemasolar central receiver system [1]

CSP technology. In this system, many mirrors, often hundreds, called heliostats, are used to reflect the sun's rays onto a receiver tower. The energy received by the tower can be transferred to a heat transfer fluid. This heat can then be used to generate steam for powering a steam turbine for electric power generation. An example of this technology is shown in Figure I.1.

CSP is of significance globally, as there is a move towards renewable energy electricity generation motivated by a concern for the environment. Among the available methods for renewable energy electricity generation, CSP emerges as a promising method due to its ability to provide power when there is no direct sunlight. This is made possible by its characteristic thermal inertia and the thermal storage for which it allows.

In the South African context, CSP is favorable since South Africa ranks high globally in terms of its available solar resource. However, due to insufficient knowledge and experience in the field of CSP, and more specifically central receiver systems, uptake of the technology into the South African electricity market is somewhat limited. To aid in relieving this issue, research into the field of CSP is essential. Hence the current work.

Heliostat Field Layouts

The design of central receiver-type CSP plants includes a design of the heliostat field. Since the heliostat field contributes significantly (up to 50%) to the overall cost of the plant, it is beneficial to ensure that the field layout is the most optimal at collecting energy from the sun. For this reason, among others, heliostat field layout optimization is an active research field.

For the heliostat fields of central receiver systems, a number of different field layouts have been proposed and tested. These layouts range from biomimetic layouts which mimic natural shapes, to rectangular grids of mirrors in straight rows. The layout chosen will influence the overall performance of the field, including the extent of blocking, shading and cosine losses—optical characteristics of the field that affect the amount of energy reaching the receiver.

The choice of a field shape and the actual placing of the heliostats, therefore, has a direct influence on the performance of the plant. Heliostat field optimization is thus crucial to plant optimization.

Objective

The objective of this book is to provide a clearer understanding of the task of heliostat field analysis and layout optimization and to present an approach in these tasks using techniques that engineers and scientists are more accustomed to. Thus, given a site and the available design resources, such personnel may be able to perform analyses and optimization of heliostat fields insightfully and intuitively. This objective is met through the demonstration of a central receiver system model adequate for optimization, a study of the optimization methods available, and by a demonstration of optimization. Multivariate gradient-based optimization in the context of heliostat placement is demonstrated with the purpose of providing clarity on how this more classical optimization method can be used in this research area.

Scope

The approach presented herein is primarily concerned with the optical efficiency of the heliostat fields of central receiver plants. A model of such plants is presented and compared with ray tracing to demonstrate its accuracy. Optimization by means of an available optimization algorithm is presented.

The information contained in this book is not a comprehensive study of all the heliostat field analyses and layout optimization methods available, nor is it a study of optimization itself or the different applicable algorithms. It is a single approach, among many, to heliostat field analysis and heliostat field layout optimization.

The study area of heliostat field layout optimization is vast. Yet, even slight probing into this area provides deep insights and invaluable learning.

Overview

This book is organized as follows. First a survey of the knowledge base is presented describing the aspects of heliostat field optimization relevant to the objective and how field layout optimization fits into the broader field of central receiver systems and renewable energy technology. A review of optimization techniques is presented as well as the common methods of heliostat field evaluation and layout optimization.

A heliostat field analysis model is presented. This is followed by a description of the validation exercises carried out to determine the accuracy of the model. Optimization is then demonstrated, and a redesign of a commercial plant, PS10 is presented. Thereafter a section is included which gives a description of the insights gained by the application of the approach presented in this book. Finally, conclusions and recommendations are presented.

CHAPTER 1

HELIOSTAT FIELDS IN CONTEXT

This chapter provides an introduction to and an overview of heliostat field layout optimization. It places heliostat field layout optimization in the context of the broader subject of renewable energy technology and describes some of the current work in this area.

1.1 Renewable Energy Technology

According to Scheer [2], the traditional energy system, which relies largely on fossil fuels, will soon be unable to sustain modern society. The author suggests that the use of fossil fuels for power generation should be replaced by renewable energy sources for the benefit of the environment, the economy and society.

Visagie and Prasad [3] state that economic growth usually happens at the expense of the environment, leading to environmental degradation, but the adoption of renewable energy technology makes it possible to benefit both the environment and the economy. The technology provides for the energy needs of the economy while causing minimal adverse effects to the natural environment. This is in contrast to conventional energy technology, such as coal-fired power stations, that negatively

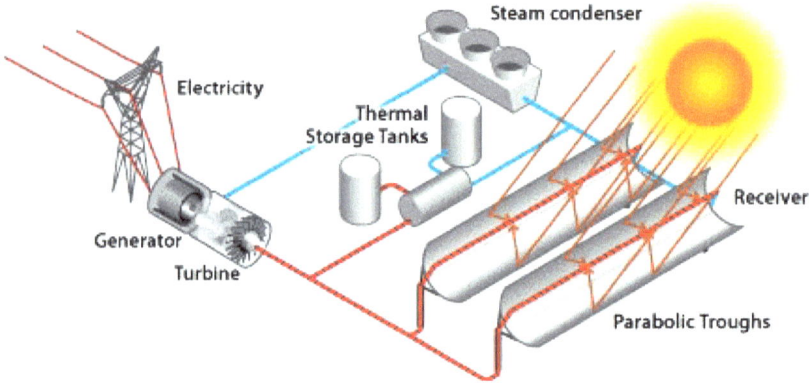

Figure 1.1: Integration of CSP collectors into a conventional power cycle [7]

affect the environment chiefly through the gases that they produce as a byproduct of operation.

Scheer [4] asserts that power generation solely by renewable energy resources is not only attainable, but necessary for the continuation of modern living. It is not certain as to when conventional fuels, such as oil and coal, will be depleted. What is known, though, is that these are finite resources and they will at some point be depleted. We may also reach a point where it is not financially feasible to extract these resources from the earth. Thus, at some point, transition to other fuels will be required to sustain modern living. For the reasons stated previously, renewable energy sources are most attractive.

Renewable energy technology makes it possible to utilize renewable energy sources such as wind, solar, hydro, biological and geothermal energy for electricity generation [4]. Stine and Geyer [5] describe the workings and applications of solar thermal energy technology—the technology used for converting radiation from the sun into usable thermal energy. The authors also describe an enhancement of this technology—CSP—which involves concentrating the solar radiation to a smaller area to achieve high temperature thermal energy.

1.2   Concentrating Solar Power

According to Pitz-Paal [6], CSP collectors can be used as the thermal energy supplier in a conventional power cycle as opposed to combustion of coal. This is illustrated in Figure 1.1. The thermal energy is supplied from collectors which concentrate solar energy to provide high-temperature thermal energy to the power block. The author mentions that CSP is most suited for centralized power production in areas that have a high solar resource.

CSP systems include linear Fresnel, parabolic trough, central receiver, and dish Stirling systems [8]. Linear Fresnel and parabolic trough systems are known as line-focus systems because they focus radiation from the sun to a line that extends along the length of the collectors [9].

Mehos [10] highlights the potential of CSP systems to contribute significantly to power generation by means of renewable energy resources. The author also mentions the ability of such a system to provide power beyond the daylight hours due to its potential for thermal storage. That is, solar energy can be utilized for electricity generation when the sun is not available. Kuravi *et al.* [8] mention that thermal energy storage is more efficient and cost-effective than mechanical, such as pumped storage schemes, and chemical storage technologies, such as batteries.

1.3 Central Receiver Systems

Schell [11] states that central receiver systems perform significantly better than other CSP systems, such as line-focus systems. Yogev *et al.* [12] mention that this is due to the high energy flux densities and, consequently, the high temperatures that may be realized by these systems. Danielli *et al.* [13] attribute the performance advantage to the dual-axis tracking system that is characteristic of heliostats, the collector mirrors in CSP systems. The author reports that this allows for a more uniform optical efficiency over the course of a year.

The line focus systems, linear Fresnel and parabolic trough, employ single-axis tracking. That is, the focusing mirrors track the sun along a single axis. The mirrors rotate east to west, or north to south only as they track the sun and reflect its light to the focus area. Dual-axis tracking systems are able to rotate both east to west and north to south, allowing for significantly lower optical losses.

1.4 Heliostat Fields

When considering both material costs and labor costs, Kolb *et al.* [14] report that the heliostat field is the "largest single capital investment" in a central receiver system. The authors further report that the greatest method for reducing the total cost of a central receiver plant significantly (in terms of capital equipment cost), is by improving the efficiency of the heliostat field. A highly efficient heliostat field requires less effective heliostat area, thus the cost of the field is minimal.

Optical efficiency of the heliostat field is determined by the layout of the heliostats, the surface of the reflective facet surface (the coating used) and the methods of keeping the reflective surface clean [14]. An improvement in any of these parameters will result in an improved heliostat field optical efficiency.

1.5 Heliostat Field Analysis

Garcia *et al.* [15] divide the methods for measuring the strength of a heliostat field into categories based on the underlying mathematical algorithm. Two categories of algorithms are presented by Garcia *et al.* [15]. The first is the Monte Carlo ray tracing algorithm upon which the codes MIRVAL and SolTRACE are built. The second category includes the algorithms based on so-called "convolution" methods. Convolution methods, as the name implies, are combinations of functions that approximate reality [15]. For this reason, the author also refers to them as approximation methods.

Shuai *et al.* [16] describe ray tracing as a method of tracking the paths that rays from the sun are most likely to follow as they interact with surfaces in the system being analyzed. The surfaces include the heliostats and the receiver. The properties of each surface, especially the reflectivity, are well-defined so that interaction between ray and surface is representative of reality. Each ray carries a certain amount of energy, state Bode and Gauché [17], and the energy intercepting a surface can be determined by summing the number of rays that hit that surface.

Garcia *et al.* [15] mention that ray tracing is more computationally expensive than the approximation methods and thus discourage the use of ray tracing techniques for optimization. Ray tracing is mainly used for optical analysis, but results of the ray tracer can be used as part of a higher-level plant analysis [18].

Garcia *et al.* [15] note that many codes are highly accurate and thus require a large amount of computational resources. Leonardi and Aguanno [19] suggest that highly accurate codes are unnecessary and that simpler, approximate methods would require less time while still providing satisfactory results. Garcia *et al.* [15] state that such simplifications are plausible since errors are usually much higher in the models of the other components—such as turbine and storage—than in the optical model.

Leonardi and Aguanno [19] describe a convolution method for accounting for all the geometrical quantities in one unique function. This function considers all the geometrical qualities of the field as optical efficiency evaluations. The efficiency can be determined at each hour and combined to determine annual efficiency.

In determining the losses experienced by a field, Noone *et al.* [20] state that blocking and shading are computationally expensive and for this reason, very few codes calculate these without introducing approximations. The author also provides a simplified model for determining blocking and shading. The heliostat face is divided into a grid, and the blocking and shading potentials are determined for each grid element.

Noone *et al.* [20] explain that the blocking and shading computational expense can be reduced by first determining the blocking and shading potential of a pair of heliostats by a proximity calculation, and then only calculating the extent of blocking if the heliostats are in close proximity. This is illustrated in Figure 1.2. Some codes neglect shading altogether, for, as Collado [21] mentions, the effect of blocking far exceeds the effect of shading. Others assume that blocking and shading rarely happens between multiple heliostats. That is, only one heliostat will block or shade another heliostat [20].

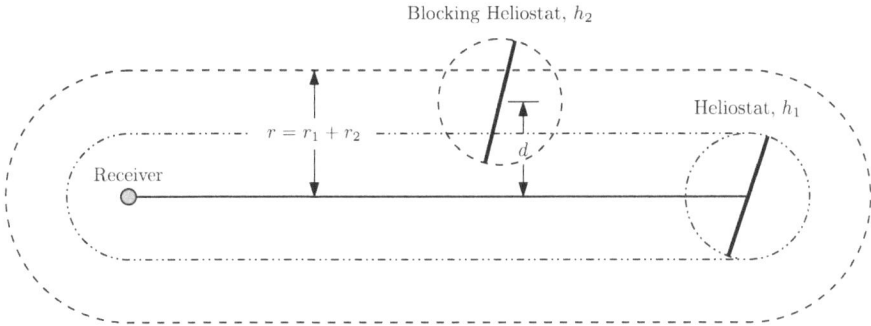

Figure 1.2: Schematic of proximity analysis [20]

Macro evaluations are also available. Among these are the "quick evaluation" pre-sented by Collado [22], and the "model for high-level decision making" by Gauché *et al.* [23]. These methods approximate the strength of a heliostat field without taking individual heliostats into consideration.

1.6   Optimization

Optimization is the process of determining the parameters that minimize an objective or fitness function [24]. A number of optimization methods exist. The methods relevant to this approach are described below.

1.6.1   Gradient Methods

Gradient methods make use of information about the gradient of the function to find the optimum [25]. Gradients can be determined by differentiation of the objective function. If the function cannot be differentiated analytically, as is the case with complex computer programs, gradients can be determined by finite difference calcu-lations or by automatic differentiation if the source code of the objective function is available [26].

Venter [25] mentions that gradient methods are able to solve problems with large numbers of design variables. The author also states that the algorithms usually do not require a lot of adjustment for the specific problem. However, they have the disadvantage of converging to local minima.

1.6.2   Modern Methods

Modern methods of optimization include the evolutionary algorithms that have be-come very prominent in the last 20 years [25]. Modern methods take their inspiration from processes and phenomena existing in nature [24]. The genetic algorithms [27], for example, mimic the evolutionary process of random change and natural selection. Particle swarm optimization mimics the social behavior and motion of a flock of birds or swarm of insects as they search for food [28].

Modern methods are able to perform global optimization, states Venter [25]; given a multimodal objective function they are less prone to converging to local minima. However, they are limited in terms of the number of design variables and the size of problem that can be solved. Furthermore the constraint-handling abilities are poor.

1.6.3   Sequential Approximate Optimization

Often in engineering design problems, the objective function that is to be minimized is not expressed in terms of the design variables. That is, the objective function may be an involved analysis or simulation, such as fluid mechanics or thermodynamic analysis, that takes in the design variables as inputs.

Nakayama *et al.* [29] mention that these analyses are usually highly time-consuming; it requires a considerable amount of computational time to obtain a value of the objective function. The authors mention sequential approximate optimization (SAO) as an optimization method that aids in relieving the computational burden by minimizing the amount of analyses of the objective function. Groenwold *et al.* [30] mention that SAO is the preferred method when computationally demanding models are used for the objective function.

To minimize the number of objective function analyses, SAO methods firstly predict the form of the objective function and construct an approximate model (or metamodel) of the function based on this prediction and then, secondly, optimize the predicted objective function. Prediction of the objective function is done using methods of computational intelligence [31]. This predicted form of the objective function is less computationally demanding.

Nakayama *et al.* [29] further state that the problem that then arises is finding a good approximation of the objective function using as little sample data as possible. Barthelemy and Haftka [32] provide an overview of the approaches that have been proposed to perform this operation.

SAO is a gradient-based method and therefore suffers from the drawback of converging to local minima. However, as with all gradient-based methods, it is efficient in terms of the number of function evaluations required to find the optimum and is able to solve problems with a large number of design variables [25].

1.7   Field Optimization Methods

Heliostat field layout optimization is generally done using one of two methods: the field growth method and the pattern method. These are described below.

1.7.1   Field Growth Method

Sánchez and Romero [33] employ what is known as the field growth method. This method starts with an empty field. Every point in the field is evaluated to find the

Figure 1.3: Field growth method procedure

best position for one heliostat to be placed, and a heliostat is assigned in this position. Then every point in the field is once again evaluated to find the best position for a second heliostat to be placed taking the previously placed heliostat into consideration, and a second heliostat is placed in this position. This process is repeated for heliostats three and four and so on until the field size is able to meet the system requirements. This procedure is illustrated in Figure 1.3.

Sánchez and Romero [33] evaluate each point in the field to determine how much energy can be collected from that point over a year if a heliostat were to be placed there. The authors call this the "yearly normalized energy surface" (YNES). A heliostat is placed at the best location. The yearly normalized energy surface is determined again, this time with the first heliostat placed taken into consideration. The second heliostat is placed at this point. This procedure is repeated until the heliostat field meets the required power output.

For the first evaluation, blocking and shading are not considered since there are no other heliostats in the field. Only once the first heliostat has been placed are the field points evaluated with blocking and shading considerations. The number of points in the field that are evaluated can be varied to improve or decrease accuracy and, consequently, computational time.

A simple search algorithm can be used, and discontinuities—such as streams, holes or restricted areas where heliostats cannot be placed—can easily be incorporated into the optimization. However, since each heliostat is to be evaluated at all possible locations, the time to determine the location of each successive heliostat rises as the optimization progresses. This is because, with each added heliostat, another blocking and shading calculation is added to the search, and these operations are the most time-consuming of all the field evaluations [20]. The search time drops again once the possible locations have diminished sufficiently.

In addition, each heliostat allocation is dependent on the preceding allocation. This leaves little space for parallelization of the optimization; it is not possible to place heliostats simultaneously. Parallelization can be employed during the search phase, though; that is, for a single heliostat placement, each possible location can be evaluated simultaneously through parallelization.

With the growth method, a feasible field can be obtained only once the optimization has been completed. This is unlike the methods that follow, which can be halted

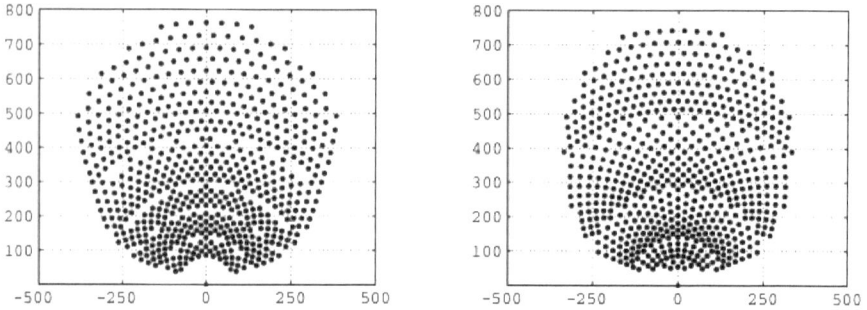

Figure 1.4: Heliostat field layout patterns [20]

at any time yet still deliver a feasible field. For this reason, the method is not suitable for large fields when adequate computing power and time are not available.

1.7.2   Pattern Method

Heliostats in a field can be arranged into elegant geometric patterns. Examples of these patterns can be seen in Figure 1.4. The patterns have certain parameters that define them. The radial stagger pattern from Stine and Geyer [5] for example, shown in Figure 1.5a, is defined by the two parameters A and R, which characterize the spacing between the heliostats. To optimize a pattern, the only variables that need to be optimized are the defining parameters.

In the case of the stagger pattern, there are only two variables that need to be optimized. Since this is a very small optimization computationally, it is simple to add a few more variables that may assist in field design. These variables could include (among other parameters) the tower height, heliostat size and the position of the first row of heliostats relative to the tower.

Several patterns are available for the pattern method. These patterns include rows, radially staggered, spirals, and the biomimetic patterns. Biomimetic patterns are patterns that mimic naturally occurring patterns, such as the phyllotaxis disc pattern employed by Noone *et al.* [20]. This pattern is shown in Figure 1.5b.

A drawback of the pattern method is that an optimized pattern does not necessarily result in an optimal field. Buck [34] has shown that improvements are possible. In the pattern method, it is not the x-y co-ordinates but the pattern parameters that are being optimized for. The x-y co-ordinates are dependent on the pattern parameters. The pattern method essentially determines the best adaptation of the pattern for the problem and not necessarily the best x-y co-ordinates for optimal plant performance.

In addition, the pattern method is not able to handle elevation variations and discontinuities within the site efficiently. To use a field optimized by the pattern method to its full potential, the site needs to be level and continuous.

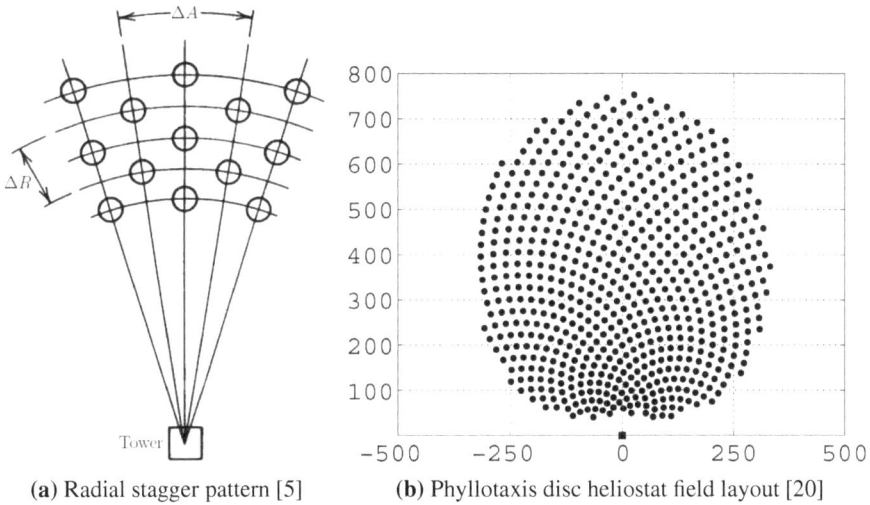

(a) Radial stagger pattern [5]   (b) Phyllotaxis disc heliostat field layout [20]

Figure 1.5: Field layout patterns

1.7.3   Other Methods

Buck [34] applies a method called "non-restrictive optimization". A field that has been optimized by a pattern method is further improved by localized gradient-based optimization. This is done by perturbing each heliostat position within a small area around the heliostat to find a better function value. If a heliostat perturbation does produce a better function value, the new location is kept. Buck achieved a 0.7% improvement in annual intercepted energy on the PS10 field, a commercial CSP plant located in Spain.

DELSOL [35], a field design tool developed by Sandia National Laboratories, employs both a field growth method and a pattern method for heliostat field layout design. Initially, individual heliostats are not taken into account. The heliostat area surrounding the tower is divided into a number of zones, and the average field performance at each zone is calculated. The zoning is shown in Figure 1.6. Once the best zones are selected, DELSOL places and optimizes a radial stagger pattern heliostat sub-field inside each zone.

So, the field growth method is used to determine what zones within the site to use and the pattern method is used to determine where individual heliostats should be placed inside the chosen zones. The zones are rated by a performance/cost ratio. Then, starting with an empty field, zone by zone is added to the heliostat field giving zones with better performance/cost ratios preference until the total power output required is reached.

With DELSOL, the optimization variables include not only the pattern parameters but also the tower height and receiver size. The main inputs for the design are (1) receiver type (2) a range of possible receiver sizes (3) a range of tower heights (4) a

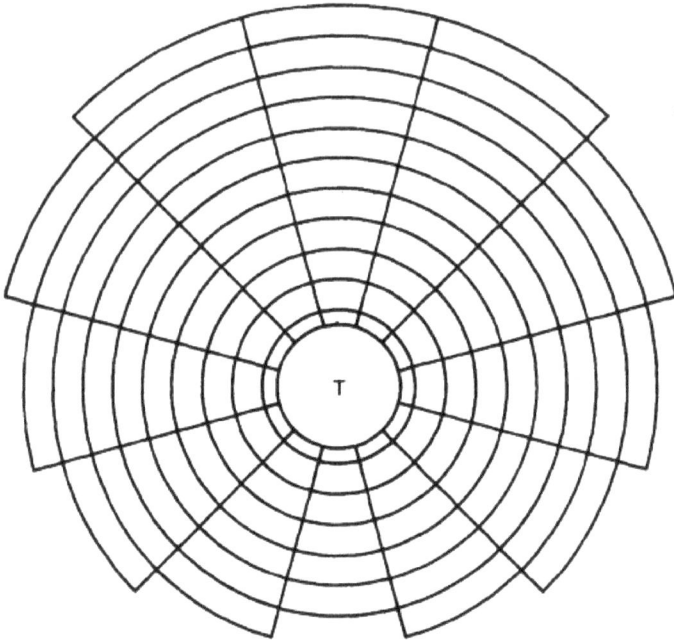

Figure 1.6: Zoning in DELSOL [35]

range of power levels and (5) flux and land constraints. Using these inputs the code generates the optimal radial stagger field layout, that is, the radial stagger layout that gives the lowest energy cost. This is done over the range of receiver sizes and tower heights. The result is an optimal radial stagger field layout with corresponding optimal tower height and receiver size on a performance/cost basis. Further optional optimization may be done by varying the heliostat density within each zone.

1.8 Field Optimization Considerations

In optimizing a field, there are a number of considerations that are prevalent in the literature. These are discussed below.

1.8.1 Objective Function

Buck [34] optimizes for maximum annual intercepted energy. He suggests, though, that maximum thermal power output of the receiver may be a promising objective too. Pitz-Paal *et al.* [36] perform an optimization where the objective is to reach maximum solar-to-chemical energy conversion efficiency of a solar thermochemical process used for producing solar fuels.

Figure 1.7: eSolar CSP power plant [37]

In their optimization, Sánchez and Romero [33] also pursue maximum annual intercepted energy as their objective. For this reason, heliostats are located according to a possible location's annual intercepted energy potential. Positions are allocated one by one, best first, until the field reaches a size that is able to meet the total power requirements.

According to Kolb *et al.* [14], one of the main considerations of central receiver systems is the overall cost of the plant and hence, the levelized cost of energy from the plant. For this reason, economic considerations may also be a useful objective in optimization. Schell [11] mentions that low-cost design and high-volume manufacturing was the main driver for the eSolar plant field layout. This plant, with its straight rows of heliostats, is shown in Figure 1.7.

1.8.2   Parameters

Buck [34] uses each co-ordinate of each heliostat as parameters in his optimization. However, he optimizes on a local basis—heliostat by heliostat—so during the optimization there are at most 2 variables at any given time.

In their optimization, Pitz-Paal *et al.* [36] take field spacing parameters, geometry of a secondary concentrator and the geometry of the reactor in their system as optimization parameters. The authors keep all other parameters fixed and subject them to a sensitivity analysis after optimization.

According to Kolb *et al.* [14], heliostat facet sizing is a largely unexplored parameter for field optimization. Research into heliostat canting by Landman and Gauché [38] suggests that the facet shape may also be a possible optimization parameter.

1.8.3   Algorithms

Pitz-Paal *et al.* [36] do a relatively small optimization and thus note that modern methods of optimization are favorable for their problem. These include the genetic algorithm, the Nelder-Mead algorithm and the Powell algorithm. Due to its poor performance, the authors find the Powell algorithm least favorable.

Collado [21] mentions that one of the problems of layout optimization is determining the initial field from which to start the optimization. The author notes that HFCAL, a software package for field optimization, starts with some unknown initial hypothetical field before optimization and suggests a method for determining this initial field, which he denotes the "preliminary design".

Pitz-Paal *et al.* [36] state that the field optimization problem is highly multimodal, making it expensive to find a global optimum. The authors suggest using a statistical method to determine a reasonably good local optimum, which can be carried out with reasonable effort.

1.8.4   Other Considerations

Aside from the strength of the heliostat field, Buck [34] mentions that there are other important aspects to be considered in field optimization. These include accessibility requirements such as roads for maintenance of mirrors, as well as area constraints such as site boundaries or restricted areas.

1.9   Software

Bode and Gauché [17] present a review of available software for heliostat field analysis and optimization. The majority of software is only available commercially. The other packages are either free to use, open source or available strictly for academic purposes.

1.10   Summary

This chapter has demonstrated that there are several methods of conducting heliostat field analysis and optimization. There are at least two different ways of determining the strength of a heliostat field, and there are numerous objectives that can be pursued. Software for determining heliostat field strength and for optimizing is available, but it is also possible to develop a model using available literature and geometric analysis.

In terms of the heliostat field layout optimization methods, there is a noted absence in the literature of a method that follows the classical approach of optimization—a method that starts with an initial layout and allows each of the heliostats to gravitate freely to an optimal location. An evaluation of the optimization techniques available is also needed. The approach used to explore some of these areas is presented in the following chapter.

CHAPTER 2

A CLASSICAL APPROACH

The purpose of this book is to demonstrate a classical approach to heliostat field analysis and layout optimization and in so doing provide improved understanding of heliostat field layout optimization. The book also seeks to demonstrate the effectiveness of multivariate gradient-based optimization in heliostat field placement. What follows is a discussion of the different components of this approach. Also described is the method that the author used in implementation of this approach.

2.1 Methodology

2.1.1 Receiver Plant Model

As part of the optimization process, an objective function is required. In the area of heliostat field optimization, this objective function should involve a field strength analysis based on a model of the plant. The objective function must calculate some characteristic of the plant that can be used to rank plants. In this approach, the commonly utilized characteristic of annual intercepted energy is used.

The intercepted energy is the energy that arrives at the receiver after optical losses have taken effect. Some of this energy is lost through convection, radiation and

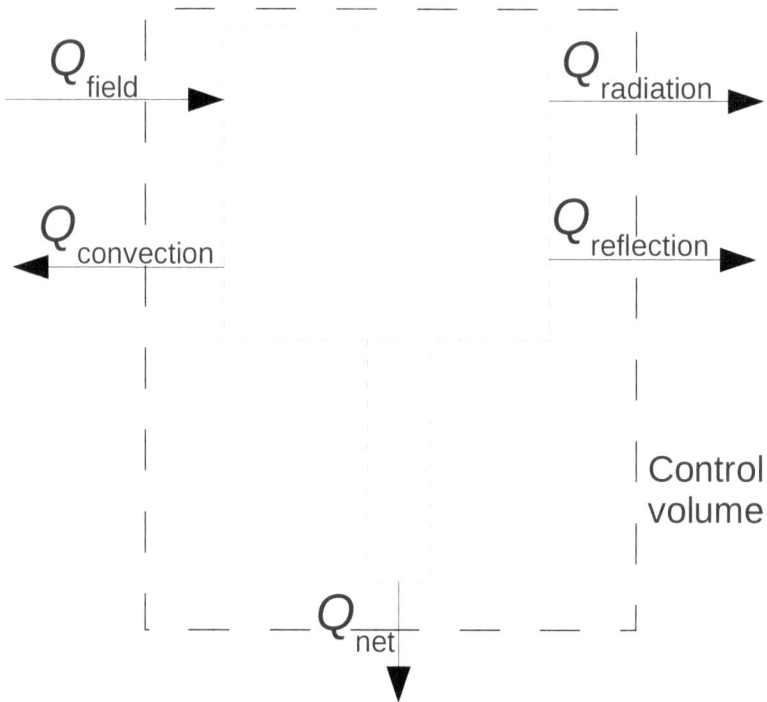

Figure 2.1: Energy balance diagram of a central receiver

reflection. The remainder can be used in the power conversion processes of the plant to generate electricity. The energy balance of the receiver is illustrated in Figure 2.1. The intercepted energy is represented in the diagram as Q_{field}, that is, the usable energy from the field.

$$Q_{net} = Q_{field} - Q_{convection} - Q_{radiation} - Q_{reflection} \qquad (2.1)$$

To assist in optimization a receiver plant model based on geometric analysis and existing models obtained in the literature is used. The model is a convolution method; that is, it is a combination of functions that approximate various features of the field, specifically, the optical efficiencies of each heliostat. These optical efficiencies are the building blocks of determining annual intercepted energy.

For testing purposes, the author compiled this receiver plant model into a computer code using the programming languages Octave [39] and Fortran [40]. Octave is an open source reproduction of the commercial software Matlab [41]. Octave was used to compile solar resource data, including sun positions, into a usable form for the Fortran program. The Fortran program contains the actual receiver plant model.

Fortran was used because of its versatility and because it ran effortlessly with the high performance computer used for the optimization. The optimization algorithm

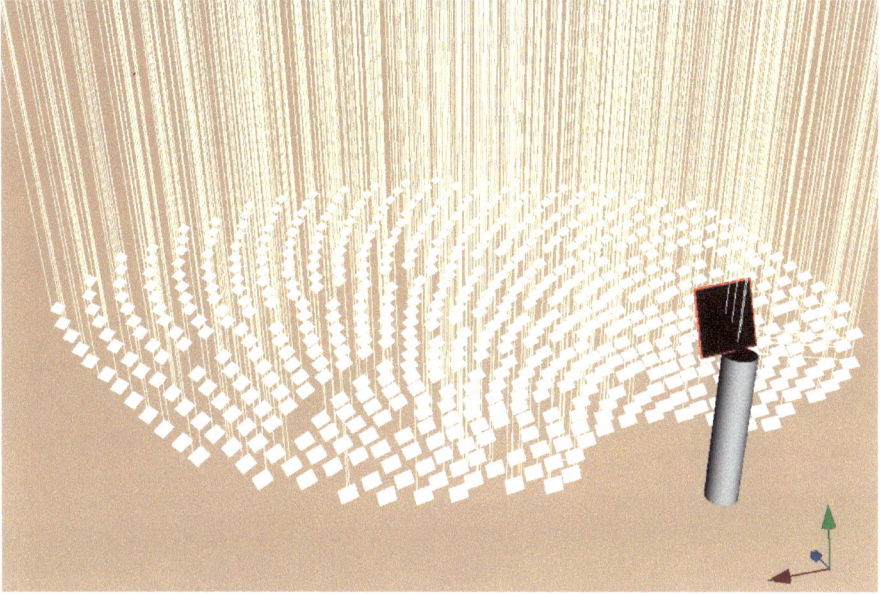

Figure 2.2: Ray tracer model of a heliostat field

was also written in Fortran. Thus it was simple to combine the receiver plant model with the optimization code.

2.1.2 Validation

To validate the receiver plant model, validated ray tracing software was utilized. Ray tracing does not need to calculate all of the efficiencies that are involved in convolution methods; most of them are implicit. This excludes atmospheric attenuation, though. Atmospheric attenuation needs to be calculated explicitly by adjusting the energy value of each ray to account for this property based on the distance that the ray travels. Figure 2.2 shows a ray tracer model of a heliostat field.

Ray tracing is highly accurate, but as mentioned in the previous chapter, due to computational intensity, though possible, it is not practical to use ray tracing methods for optimization. However, since it is based on different principles, racing tracing is useful for evaluating the accuracy of a convolution method model. The author, therefore used ray tracing to determine the accuracy of the plant model presented herein.

A number of test cases were carried out using the developed plant model. These test cases were reproduced with the ray tracing software. The results of the ray tracing tools were compared with the results of the plant model.

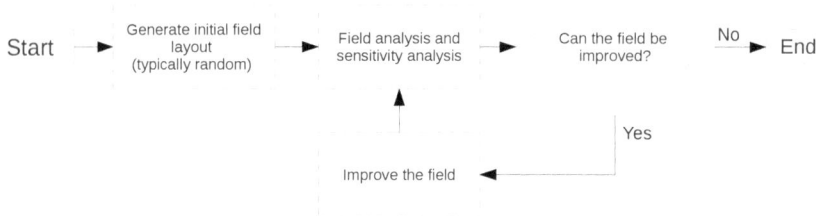

Figure 2.3: The free variable method of heliostat field layout optimization

2.1.3 Optimization

As indicated in the previous chapter, the current literature in this area indicated the prevalent use of two optimization methods: the pattern method, done by using field spacing parameters as optimization parameters, and the field growth method, done by allocating heliostats one by one to the best positions in a field. These methods were studied briefly. However, it was discovered that a third method of heliostat field optimization exists. The current author calls this method the "free variable method."

The free variable method follows a more classical approach to optimization. Each heliostat in the field is assigned a location. Then, through successive iterations of sensitivity computations and field analyses, heliostats are allowed to gravitate freely to points that produce an optimal overall field performance. Figure 2.3 shows a basic illustration of this method. This method is uncommon in heliostat field optimization. Little if any literature exists on the subject.

To perform this method of optimization successfully, an appropriate optimization algorithm was required. Unlike the growth method and the pattern method, the choice of an optimization algorithm for the free variable method is not straightforward. The reason for this is the large number of variables and constraints associated with the problem. It was found that this method required an algorithm capable of large scale optimization where the number variables is large and the number of constraints outnumber the variables.

Furthermore, the analysis method used for the objective function was a highly involved mathematical computer script. For this reason, sequential approximate optimization seemed attractive.

The author used a previously developed algorithm for this optimization operation: the SAOi algorithm (see Appendix C). This algorithm is written in Fortran and makes provision for any objective function written in Fortran.

The receiver plant model developed was combined with the SAOi code. Some initial tests were performed to ensure that the receiver plant model and optimization code were working together properly. Thereafter, the optimization was prepared for running on a high performance computer. A number of optimizations were then performed.

2.2 Limitations

The validation technique used was a comparison of the model developed with previously developed and validated models. This does provide a picture of the validity of the model developed in this approach adequate for the purposes of this approach. However, it may be lacking in terms of providing an accurate picture of how the model compares with reality. The data collected in the validation shows some differences with the models. It is not entirely clear whether, and to what extent, these differences indicate that the model is more representative of reality or less.

2.3 Summary

The methodology of the classical approach purported in this book has been presented. The plant model used as an objective function for the optimization process is a convolution method developed from approximation models available in the literature. The optimization algorithm used is based on sequential approximate optimization.

This approach of heliostat field layout optimization the author calls the free variable method—a method that seems to be absent in the literature. This optimization method appears to be powerful but may pose a challenge in terms of computational requirements.

The objective function to be used is based on an approximation model of the field and determines the annual intercepted energy. This model is the subject of the next chapter.

CHAPTER 3

RECEIVER PLANT MODEL

A method of evaluating the strength of a field is required as part of the field optimization process. For this, a model of the plant is required. This section describes such a model. The model calculates the amount of energy that can be collected by a heliostat field over a year. It is not an exact calculation; it is an approximation. Yet, it is adequately accurate for the purpose of this approach as will be demonstrated. It is an adaption of the function used by Leonardi and Aguanno [19].

3.1 Intercepted Energy

The sun emits a large amount of energy. However, only a fraction of this energy reaches the earth. Also, since the earth is rotating, any collecting device situated on earth can only harvest energy from the sun for a portion of the day. Usable solar radiation is further reduced by the presence of the atmosphere and by the inefficiencies of the collecting device [5].

In central receiver systems, the collector consists of the array of heliostats in the field and the central receiver to which the sunlight is reflected. The amount of energy that the heliostat field delivers to the receiver is known as the intercepted energy. In field analysis, the effectiveness of the heliostat field in harvesting the energy from the sun is evaluated. The analysis, therefore, requires a model of the field characteristics.

The intercepted energy at each hour of the year can be determined by multiplying the effective area of each heliostat by the available direct normal irradiation (DNI) at that hour [19]. The effective area can be expressed as the total area of the heliostats multiplied by the optical efficiency of each heliostat. The following equation, adapted from Leonardi and Aguanno [19], illustrates this:

$$I = A \sum_{h=1}^{8760} \text{DNI}_h \left(\sum_{i=1}^{m} \eta_{c_{i,h}} \eta_{a_i} \eta_{sp_i} \eta_{b_{i,h}} \eta_{s_{i,h}} \right) \tag{3.1}$$

I is the intercepted energy and A is the reflective surface area of a single heliostat. The equation assumes that all heliostats have the same areas. The subscript, h, represents the hour being considered and i is the heliostat number. The cosine, atmospheric attenuation, spillage, blocking and shading efficiencies are represented by each of the η terms respectively. So $\eta_{c_{i,h}}$, for example, is the cosine efficiency of heliostat i at hour h.

The optical efficiencies of each heliostat, which include blocking, shading, cosine, attenuation and spillage, can be modeled by geometric analysis, taking into account the x and y co-ordinates of the heliostat within the field. The following sections describe how this is done. As previously mentioned, each of the functions that make up the model are approximations and not exact calculations.

Because the intended use of this model is heliostat field layout optimization, it is necessary to express the different losses of the heliostats in terms of each heliostat's location in the field. From iteration to iteration of the optimization process, the location of the heliostats will change albeit by minute amounts. Thus, the model must be generalized such that each of the losses can be calculated based on individual heliostat locations.

3.2 Sun Vector

To utilize the solar resource for solar powered electricity generation, it is important to account for the changes in solar resource caused by the sun's apparent motion through the sky [5]. Cosine, blocking and shading efficiencies are all dependent on the orientation of the heliostats. The orientation of each heliostat, in turn, is dependent on the sun vector—the vector pointing from the heliostat to the sun. This vector gives an indication of the angle of the sun's rays with respect to a horizontal surface at the location of the heliostat.

For the sun's rays to be reflected to the target, the heliostat normal must bisect the sun vector and the target vector—the vector from the heliostat to the target. To determine the sun vector, the zenith angle, θ_z, and the solar azimuth angle, γ_s, need to be calculated. These angles are indicated in Figure 3.1.

The sun vector, as a unit vector, for each hour of evaluation is determined using the following procedure adapted from Duffie and Beckman [42]:

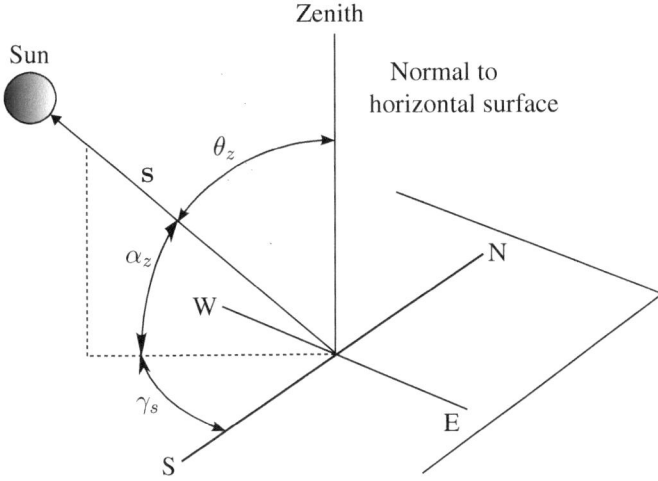

Figure 3.1: Sun vector with associated sun angles

The solar time is determined as follows:

$$\text{Solar time} = \text{Standard time} + [4(L_{loc} - L_{st}) + E]/60 \qquad (3.2)$$

where L_{st} is the longitude on which the time zone is based at the location and L_{loc} is the longitude of the actual location. Standard time and solar time are in units of hours. E, in units of minutes, is the equation of time given by the following relation:

$$E = 229.2(0.000075 + 0.001868 \cos B - 0.032077 \sin B$$
$$- 0.014615 \cos 2B - 0.04089 \sin 2B) \qquad (3.3)$$

where B is an angle corresponding to the day of the year. B is calculated as follows:

$$B = (n - 1)\frac{360}{365} \qquad (3.4)$$

and n is the day of the year.

Next, the hour angle is calculated. The hour angle gives an indication of the position (east or west) of the sun with respect to the local meridian and is calculated as follows:

$$\omega = [(\text{Solar time})/24 - 0.5] \times 360 \qquad (3.5)$$

The angle of declination is then calculated. The declination angle is the sun's angular position at solar noon with respect to the plane made by the equator. It is determined, in units of radians, as follows:

$$\delta = 0.006918 - 0.399912 \cos(B) + 0.070257 \sin(B)$$
$$- 0.006758 \cos(2B) + 0.000907 \sin(2B)$$
$$- 0.002679 \cos(3B) + 0.00148 \sin(3B) \qquad (3.6)$$

The zenith angle can now be calculated. The zenith angle is indicated in Figure 3.1 and is the angle between the sun vector and the vertical or zenith. The sun vector is the line of incidence of the sun's rays. By making use of the previously calculated parameters, the zenith angle is calculated:

$$\theta_z = \cos^{-1}[\cos(\phi) \times \cos(\delta) \times \cos(\omega) + \sin(\phi) \times \sin(\delta)] \tag{3.7}$$

where ϕ is the latitude of the location with respect to the equation, with north positive and south negative.

The zenith angle is used to calculate the solar altitude angle α_z indicated in Figure 3.1. Since θ_z and α_z are complementary angles, the calculation is simply the following:

$$\alpha_z = 90° - \theta_z \tag{3.8}$$

The solar azimuth angle, γ_s, is shown in Figure 3.1 and is is calculated thus:

$$\gamma_s = \text{sign}(\omega) \left| \cos^{-1} \left(\frac{\cos \theta_z \sin \phi - \sin \delta}{\sin \theta_z \cos \phi} \right) \right| \tag{3.9}$$

Finally, the three components of the sun vector are determined:

$$\mathbf{s} = [s_E, s_N, s_z]^T \tag{3.10}$$

The subscripts E, N and z refer to directions east, north and zenith, respectively. These components are determined as follows:

$$s_E = \cos(\alpha_z) \times -\sin(\gamma_s)$$
$$s_N = \cos(\alpha_z) \times -\cos(\gamma_s) \tag{3.11}$$
$$s_z = \sin(\alpha_z)$$

As mentioned, the sun vector is to be calculated for each hour of evaluation; thus it is referred to using the subscript h indicating the hour: s_h. The sun vector is assumed to be the same over the entire heliostat field at any moment. A sample calculation of the sun vector is included in Appendix A.2.

3.3 Target Vector

The target vector is the vector that points from the heliostat to the tower. The target vector is defined as follows:

$$\mathbf{T}_i = \begin{bmatrix} x_T - x_i \\ y_T - y_i \\ z_T - z_i \end{bmatrix} \tag{3.12}$$

where (x_T, y_T, z_T) are the co-ordinates of the target and (x_i, y_i, z_i) are the co-ordinates of heliostat i. This vector is unitized by dividing each component of the vector by the magnitude of the vector:

$$\mathbf{t}_i = \frac{\mathbf{T}_i}{\|\mathbf{T}_i\|} \tag{3.13}$$

A sample calculation of the target vector is included in Appendix A.3.

3.4 Heliostat Normal

The heliostat normal is calculated by adding the target vector to the sun vector. This gives a resultant vector that bisects the angle between the sun vector and the target vector allowing for reflection. The resultant vector is unitized by dividing each component of the vector by the magnitude of the vector:

$$\mathbf{N}_{i,h} = \mathbf{s}_h + \mathbf{t}_i \tag{3.14}$$

$$\mathbf{n}_{i,h} = \frac{\mathbf{N}_{i,h}}{\|\mathbf{N}_{i,h}\|} \tag{3.15}$$

A sample calculation of the heliostat normal is included in Appendix A.4.

3.5 Cosine Efficiency

Maximum intercepted energy occurs when a collector is perpendicular to rays from the sun. Any deviation from this position results in a reduced intercepted energy proportional to the cosine of the angle of deviation. This is known as the cosine effect. The cosine effect is illustrated in Figure 3.2.

Using the cosine effect, an associated cosine efficiency can be calculated using the Law of Reflection as stated by Noone *et al.* [20]:

$$\eta_{c_{i,h}} = \mathbf{s}_h \cdot \mathbf{n}_{i,h} \text{ (dot product)} \tag{3.16}$$

where $\mathbf{s}_h$ is the sun vector at hour h, and $\mathbf{n}_{i,h}$ is the normal vector of heliostat i at hour h. A sample calculation of the cosine efficiency is included in Appendix A.5.

3.6 Attenuation Efficiency

As light travels through the atmosphere, molecules of air and aerosols in the atmosphere cause some of the light to be scattered. This reduces the amount of energy that can be collected by a collector intercepting the radiation from a radiation source. This effect is called atmospheric attenuation. The extent of atmospheric attenuation is dependent on the distance through which the radiation travels. This is also true in the case of light emitted from the sun; energy is scattered by particles in the atmosphere as the light travels towards a heliostat.

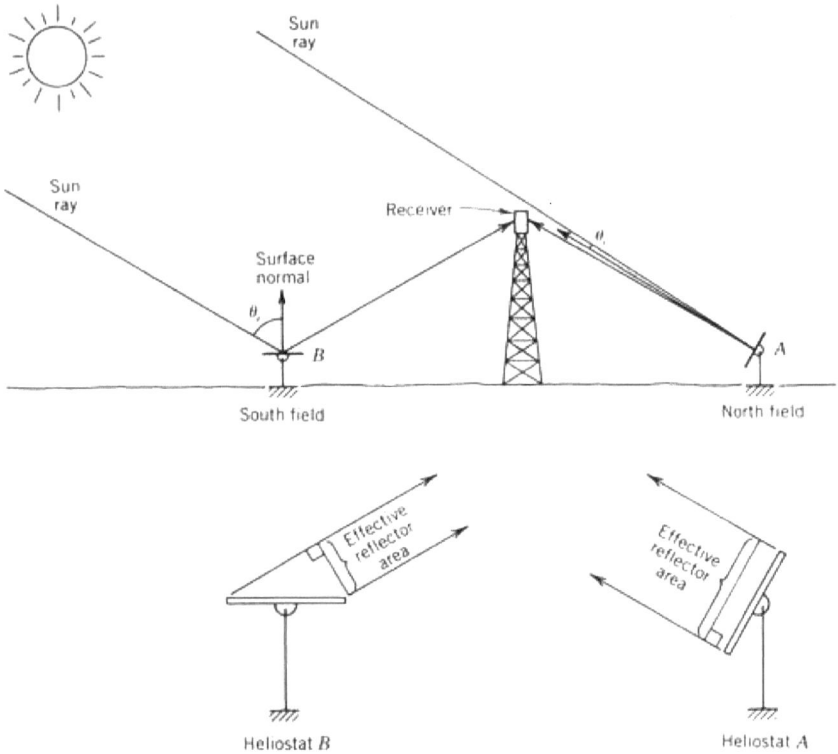

Figure 3.2: Illustration of the cosine effect [5]

The resultant energy that is intercepted by a heliostat, after the effect of atmospheric attenuation has been considered, is known as the resultant transmitted energy. The resultant transmitted energy, as a proportion of the the total energy, can be referred to as the attenuation efficiency. Using the effect of atmospheric attenuation, an expression for attenuation efficiency can be formulated. Attenuation efficiency is thus calculated using the following relation from Noone et al. [20]:

$$\eta_{a_i} = 0.99321 - 0.0001176 \cdot d + 1.97 \times 10^{-8} \cdot d_T^2 \qquad (3.17)$$

where d is the distance of the heliostat to the target, illustrated in Figure 3.3 and calculated as:

$$d = \|(x_i, y_i, z_i) - (x_T, y_T, z_T)\| \qquad (3.18)$$

and (x_T, y_T, z_T) are the co-ordinates of the target.

A sample calculation of the attenuation efficiency is included in Appendix A.6.

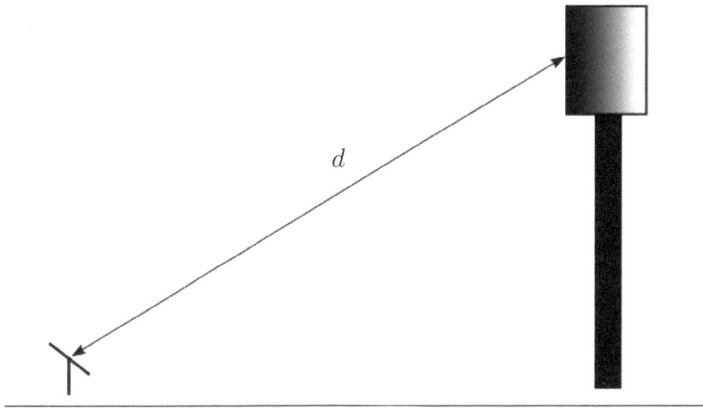

Figure 3.3: Illustration of distance used in attenuation efficiency calculation

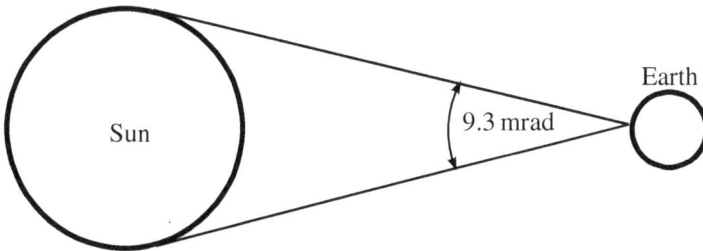

Figure 3.4: Angle subtended by the sun

3.7 Spillage Efficiency

Heliostats reflect radiation from the sun to the central receiver. The reflected radiation that falls within the perimeter of the collector apperture is used for power generation. The reflected radiation that falls outside of this perimeter is lost. This effect is known as spillage.

Each heliostat casts an image onto the receiver. Spillage efficiency is determined by approximating how much larger (if at all) the image is than the receiver. This calculation is done differently depending on the type of receiver being considered. Two receiver types are considered here: external cylindrical and flat receivers.

The rays of sunlight coming from the sun are not parallel. The rays emitted from the sun subtend an angle of approximately 9.3 mrad [5], as is illustrated in Figure 3.4. Thus, when the suns rays hit a flat heliostat and are reflected, the image spreads outwards by this angle. This is shown in Figure 3.5. This outward spread is the minimum spread that will be caused by a heliostat. The spread angle also increases due to heliostat imperfections and slope errors.

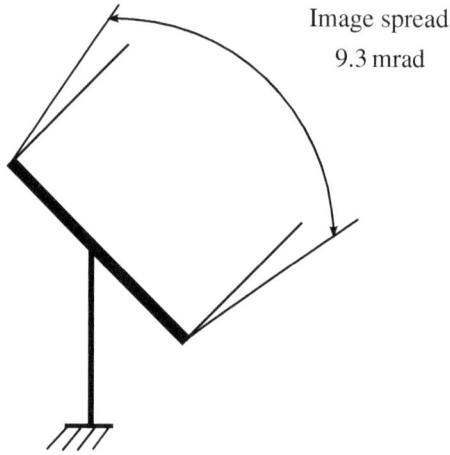

Figure 3.5: Reflected image of the sun from a heliostat

3.7.1   External Cylindrical Receivers

The spillage efficiency for external cylindrical receivers is calculated as follows:

The distance, d, from the heliostat to the target is determined. It is the same as is calculated for the atmospheric attenuation calculation. The image size is determined using the following equation, a modification of the arc length equation:

$$D_{\text{image}} = d\lambda + w \tag{3.19}$$

where D_{image} is the diameter of the reflected image at d, the distance of the heliostat from the receiver determined in the previous step. The angle, λ, is the angle which the sun subtends when viewed form the earth (Stine and Geyer [5] give this angle as 9.3 mrad) and w is the largest dimension of the heliostat (width or height).

The size of the vertical axis of the elliptical image cast onto the receiver is then calculated. By using the equation in the previous step to calculate the image size at the receiver, it is assumed that the image would be a circular sun disc if the target was normal to the target vector. Because the target is not normal to the target vector, an ellipse is formed on the target.

For a cylindrical receiver, the ellipse will have its two axes, horizontal and vertical, aligned with the horizontal and vertical of the receiver. The horizontal axis of the ellipse will be equal in length to the image size, D_{image}, at this point. The vertical axis will be lengthened somewhat based on the distance of the heliostat from the tower.

The lengthening is determined by dividing the image size, D_{image}, by the sinusoid of the angle between the target vector and the receiver. That is,

$$L_v = \frac{D_{\text{image}}}{\sin \alpha} \tag{3.20}$$

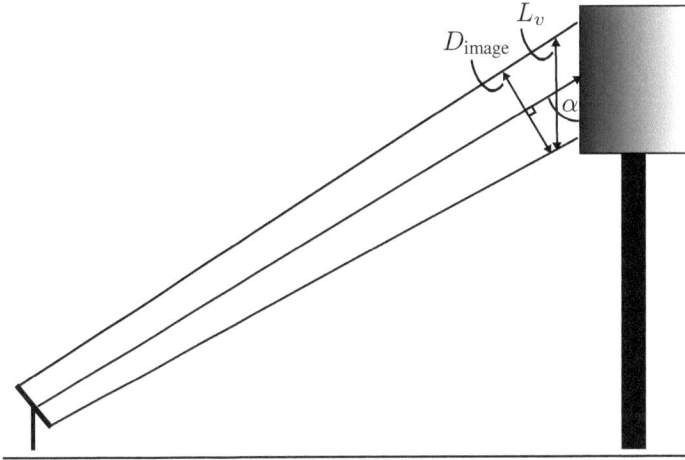

Figure 3.6: Calculation of vertical length of image

where L_v is the length of the vertical axis of the elliptical image cast on the receiver. The angle α is the angle at which the image is cast. This angle is determined as follows by the following equation:

$$\alpha = \sin^{-1} \frac{d_{xy}}{d} \qquad (3.21)$$

where d_{xy} is the distance, in the xy-plane (the ground) from the heliostat to the receiver. This applies only to an external cylindrical receiver since each heliostat effectively sees a rectangular target.

The image is compared with the receiver dimensions to determine how much of the image is spilled. If either the length of the vertical axis, L_v, or the length of the horizontal axis, D_{image}, of the elliptical image is larger than the vertical and horizontal dimensions of the receiver, respectively, the size of area outside the receiver area is calculated. This is the size of the spilled area and is calculated as follows:

The total area of the actual image is calculated using the standard equation for determining the area of an ellipse:

$$A_{\text{total}} = \frac{\pi}{4} L_v D_{\text{image}} \qquad (3.22)$$

The difference in length between the receiver width and the image diameter, D_{image}, is calculated. This is multiplied by the image vertical length, L_v, to give a representative rectangle of the spilled area. This is illustrated in Figure 3.7. Similarly, the difference in length between the receiver height and the image vertical length, L_v, is calculated. This value is multiplied by the image diameter, D_{image}, to give a representative rectangle of the spilled area.

This rectangular area in each case is then divided by a factor of 1.284 to give the actual area outside of the receiver area. The area of a rectangle is a factor of $4/\pi$, or

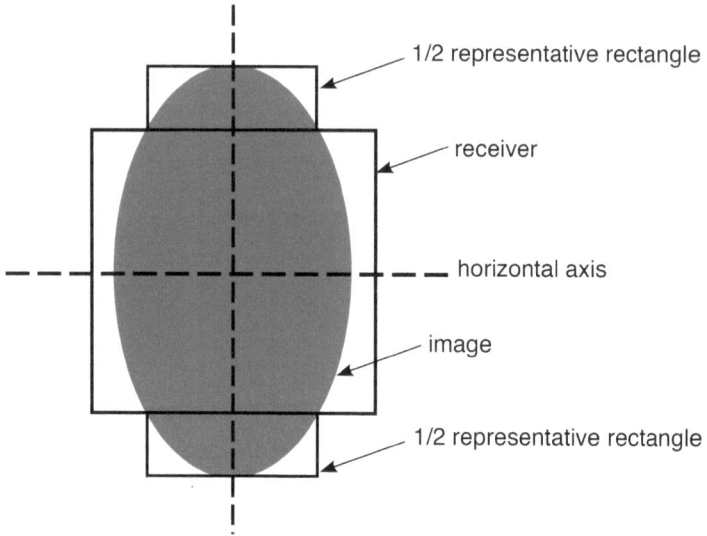

Figure 3.7: Representative rectangle of image spilled in vertical direction

1.273, larger than an ellipse of the same dimensions. The factor of 1.284 used in this model is slightly larger. This accounts well for the distortion of the ellipse within the representative rectangle. It was determined empirically using an example case where the image area, receiver area and heliostat locations were known.

The total ineffective area then is the sum of the ineffective area in the horizontal and vertical directions:

$$
\begin{aligned}
A_{\text{ineffective}} &= A_{\text{ineffective}_v} + A_{\text{ineffective}_h} \\
&= \frac{(L_v - H_{\text{Receiver}}) \cdot D_{\text{image}}}{1.284} + \frac{(D_{\text{image}} - D_{\text{Receiver}}) \cdot L_v}{1.284}
\end{aligned} \tag{3.23}
$$

where H_{Receiver} and D_{Receiver} are the height and the diameter, respectively, of the receiver.

This then gives a value for the area of the image that is ineffective. The effective image area is:

$$
A_{\text{effective}} = A_{\text{total}} - A_{\text{ineffective}} \tag{3.24}
$$

Finally, the effective area is compared with the total area to give a value for the spillage efficiency:

$$
\eta_{sp} = \frac{A_{\text{effective}}}{A_{\text{total}}} \tag{3.25}
$$

A sample calculation of the spillage efficiency calculated for a cylindrical receiver is included in Appendix A.7.

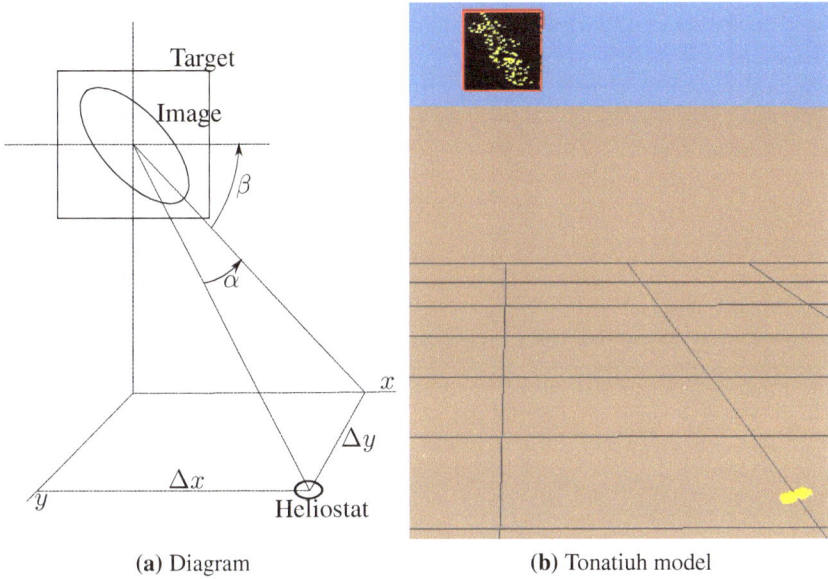

(a) Diagram   (b) Tonatiuh model

Figure 3.8: Image cast onto a flat receiver

3.7.2 Flat Receivers

The spillage efficiency for flat receivers is calculated in the same way that it is calculated for external cylindrical receivers with one exception: the angle, α, that the reflected sunlight makes with the receiver is calculated differently. This is illustrated in Figure 3.8a. It is also illustrated by the photon map of the model created in Tonatiuh [43] in Figure 3.8b. Since a flat receiver is effectively a rectangular target facing one direction, α is calculated as follows:

$$\alpha = \sin^{-1} \frac{\Delta y}{d} \tag{3.26}$$

where Δy is the distance from heliostat to receiver in the direction that the receiver is facing. As defined above, d is the direct distance from the heliostat to receiver.

A negative value for α would mean that the heliostat is behind the receiver. For the flat receiver this means that no sunlight reaches the receiver and so a negative value for α will produce a spillage efficiency of zero.

The image on the receiver will once again be an ellipse. However, the horizontal and vertical axes of the ellipse will be slanted at an angle β, shown in Figure 3.8a, from the horizontal and vertical axes of the receiver. This angle can be calculated as follows:

$$\beta = \tan^{-1} \frac{\Delta x}{\Delta z} \tag{3.27}$$

where Δx is the distance from heliostat to the vertical axis of the receiver in the direction parallel to the face of the receiver, and Δz is the height of the center of receiver above the heliostat center (i.e., the distance from the heliostat to the horizontal axis of the receiver in the direction parallel to the face of the receiver).

Using the calculated value of α, the diameter, D_{image}, vertical length, L_v, and all the areas are calculated as they are for external cylindrical receivers. Spillage efficiency is then calculated also using the total and effective areas as it is for the external cylindrical receiver.

3.8   Blocking Efficiency

Blocking occurs when something lies in the path of the heliostat and the target [15]. The blocking could be caused by another heliostat or by the pylon upon which the heliostat is mounted. Here, only blocking by heliostats will be considered. The method used here for determining the extent of blocking (as well as shading in the next section) is based on the discretization concept used by Noone *et al.* [20].

To calculate how much blocking the heliostat under consideration will experience, two questions are asked:

1. Is the potentially blocking heliostat close enough to the target vector line of the heliostat under consideration to cause blocking?

2. Is the potentially blocking heliostat closer to the tower than the heliostat under consideration?

If the answer to either of these questions is negative, blocking is not calculated and the blocking efficiency for the heliostat under consideration is set to unity.

From Stewart [44], the equation of a line in three dimensional space may be expressed as follows:

$$\mathbf{r} = \mathbf{r}_0 + t\mathbf{v} \tag{3.28}$$

where $\mathbf{v}$ is a vector parallel to the line and $\mathbf{r}_0$ is a vector from the origin to any point on the line. At this point, the scalar parameter t is equal to zero. Any value of t greater than zero defines a point on the line in the direction of the vector. Any value of t less than zero defines a point on the line in the opposite direction of the vector. This is illustrated in Figure 3.9a.

Also, the shortest distance, d, from any point, $\mathbf{p} = (x_p, y_p, z_p)$, in space to the line may be expressed as follows:

$$d = \frac{|\mathbf{a} \times \mathbf{b}|}{|\mathbf{a}|} \tag{3.29}$$

where $\mathbf{a}$ is a vector parallel to the line (which may be $\mathbf{v}$), and $\mathbf{b}$ is a vector from the point where the scalar parameter, t, is equal to zero to the point, $\mathbf{p}$. These parameters are shown in Figure 3.9b. Since the shortest distance is being found, there is an associated connecting line extending from the point to the original line perpendicular to the original line. This line intersects the original line at a certain value of the

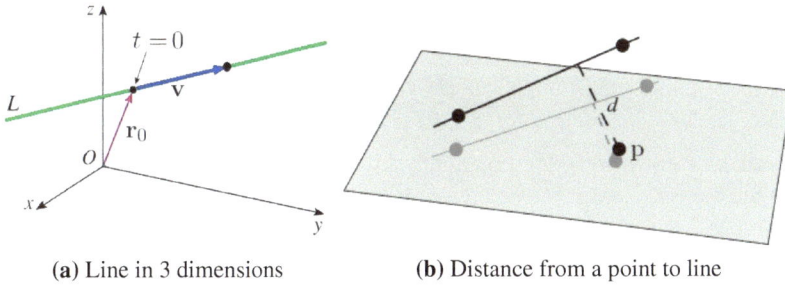

(a) Line in 3 dimensions   **(b)** Distance from a point to line

Figure 3.9: Blocking and shading geometric considerations [45]

parameter, t, given as follows from Weisstein [45]:

$$t = \frac{\mathbf{b} \cdot \mathbf{v}}{|\mathbf{v}|^2} \tag{3.30}$$

with $\mathbf{b}$ and $\mathbf{v}$ as defined above. This expression is found by differentiating an expression of the distance from the point to the line and setting it equal to zero.

These expressions are used as follows to answer the two questions stated above with regards to whether or not blocking will occur: A co-ordinate system is defined such that the center of the heliostat under consideration is at its origin. The equation of the line from the heliostat to the target can be formed. Since the heliostat center is at the origin, the equation may be expressed as:

$$\mathbf{r} = t\mathbf{T} \tag{3.31}$$

where $\mathbf{T}$ is the vector from the heliostat to the target. $\mathbf{T}$ is found using the co-ordinates of the target relative to the center of the heliostat under consideration. At the origin, or the center of the heliostat, the scalar parameter, t, is equal to zero. The shortest distance, d, from a potentially blocking heliostat to this line is determined by applying equation 3.29 as follows:

$$d = \frac{|\mathbf{T} \times \mathbf{R}|}{|\mathbf{T}|} \tag{3.32}$$

where $\mathbf{R}$ is the vector from the center of the heliostat under consideration to the center of the potentially blocking heliostat. If this distance is greater than some critical distance, d_c, blocking will not occur. This distance is taken to be the length of the diagonal of the heliostat:

$$d_c = (H_w^2 + H_h^2)^{\frac{1}{2}} \tag{3.33}$$

where H_w and H_h are the width and height of the heliostat respectively. If d is less than d_c, the answer to the first question is positive, and the next question must be evaluated.

The scalar parameter, t, associated with the co-ordinates of the potentially blocking heliostat and its shortest line of connection, is determined by applying equation 3.30 as follows:

$$t = \frac{\mathbf{R} \cdot \mathbf{T}}{|\mathbf{T}|^2} \tag{3.34}$$

If t is less than or equal to 0, the potentially blocking heliostat is not closer to the tower than the heliostat under consideration; it is behind or next to the heliostat under consideration and cannot block it. If t is greater than 0, blocking will occur.

Once it has been confirmed that blocking will occur, the extent of blocking is determined. To determine the extent of blocking, two assumptions are made:

1. A heliostat close enough to cause blocking has the same orientation as the heliostat it is blocking.

2. The geometry of the image being blocked is the same as the geometry of the heliostat from which it is reflected.

The extent of blocking is determined by an adaptation of the discretization method as described by Noone *et al.* [20]. To do this, firstly 9 nodal points are mapped out evenly over the face of the heliostat being blocked. The number of points used will determine the accuracy of the model as well as the computational effort required to do the calculation. A larger number of points will improve the model accuracy but will also increase computational expense.

Blocking and shading are the most computationally demanding calculations of all the calculations necessary in heliostat field analysis. It is vital to ensure that these computations are done only as necessary. As will be demonstrated in a subsequent section, the number of points used here results in a model sufficiently accurate for the purpose of this approach.

The points are separated along the height of the heliostat by a distance δ_h calculated as:

$$\delta_h = \frac{H_h}{3} \tag{3.35}$$

where H_h is the height of the heliostat, and along the width of the heliostat by a distance δ_w calculated as:

$$\delta_w = \frac{H_w}{3} \tag{3.36}$$

where H_w is the width of the heliostat. A schematic of the discretization is show in Figure 3.10.

The central node is at the center of the heliostat and will have the co-ordinates $(0, 0, 0)$ in the local co-ordinate system. All the other points are assigned 3-dimensional co-ordinates with respect to the central node.

Next, 9 points are mapped out evenly over the blocking heliostat. Each of the points are assigned 3-dimensional co-ordinates with respect to the $(0, 0, 0)$ point on the heliostat being blocked.

A line is projected from each of the points on the heliostat being blocked along a line parallel to the target vector extending from the point $(0, 0, 0)$ to the target. For

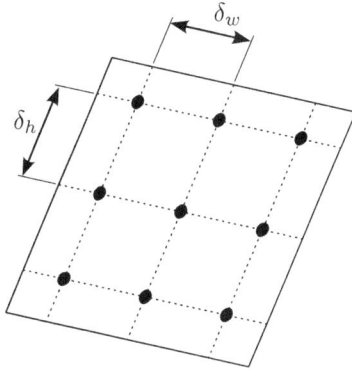

Figure 3.10: Schematic of discretization of heliostat face

canted heliostats, the lines from each point can be constructed using a unique target vector from that point to the target.

The shortest distance from each of the points on the blocking heliostat to each of the lines extending from the heliostat being blocked is determined. If the line comes within a certain critical distance, d'_c, of any of the points, the line has intersected the interior of the blocking heliostat. The critical distance used here is:

$$d'_c = \left(\sqrt{\delta_h^2 + \delta_w^2} \right) /2 \qquad (3.37)$$

The number of lines that have intersected the interior of the blocking heliostat are added up and the blocking efficiency is then determined as follows:

$$\eta_b = 1 - \frac{\text{Number of intersecting lines}}{9} \qquad (3.38)$$

since 9 is the total number of lines. A sample calculation of the blocking efficiency is included in Appendix A.8.

3.9 Shading Efficiency

Shading occurs when something lies in the path of the sun's rays and the heliostat [15]. To calculate how much shading the heliostat under consideration will experience two questions are again asked:

1. Is the potentially shading heliostat close enough to the vector line connecting the sun and the heliostat under consideration to cause shading?

2. Is the potentially shading heliostat closer to the sun than the heliostat under consideration?

If the answer to either of these questions is negative, shading is not calculated and the shading efficiency for the heliostat under consideration is set to unity.

To determine shading efficiency, the same expressions as those used in determining blocking efficiency are employed with slight adaptation. A co-ordinate system is defined such that the center of the heliostat under consideration is at its origin. The equation of the line from the heliostat to the sun can be formed. Since the heliostat centre is at the origin, the equation may be expressed as follows:

$$\mathbf{r} = t\mathbf{S} \tag{3.39}$$

where $\mathbf{S}$ is the vector from the heliostat to the sun. $\mathbf{S}$ is determined in the solar resource model. At the origin, or the center of the heliostat, the scalar parameter, t, is equal to zero.

The shortest distance, d, from a potentially shading heliostat to this line is determined by applying equation 3.29 as follows:

$$d = \frac{|\mathbf{S} \times \mathbf{R}|}{|\mathbf{S}|} \tag{3.40}$$

where $\mathbf{R}$ is the vector from the center of the potentially shading heliostat to the center of the heliostat under consideration. If this distance is greater than the critical distance, d_c (defined in the previous section), shading will not occur. If d is less than d_c, the answer to the first question is positive and the next question must be evaluated.

The scalar parameter, t, associated with the co-ordinates of the potentially shading heliostat and its shortest line of connection, is determined by applying equation 3.30 as follows:

$$t = \frac{\mathbf{R} \cdot \mathbf{S}}{|\mathbf{S}|^2} \tag{3.41}$$

If t is less than or equal to 0, the potentially shading heliostat is not closer to the sun than the heliostat under consideration; it is behind or next to the heliostat under consideration (in relation to the sun) and cannot shade it. If t is greater than 0, shading will occur, the extent of which is determined in the next step.

In calculating the extent of shading, only one assumption is necessary:

1. A heliostat close enough to cause shading has the same orientation as the heliostat it is shading.

Using the same discretization method as is used for blocking, the number of lines intersecting the potentially shading heliostat is determined. Shading efficiency is then determined as:

$$\eta_s = 1 - \frac{\text{Number of intersecting lines}}{9} \tag{3.42}$$

3.10 Topography

The center of each heliostat has co-ordinates (x, y, z). These co-ordinates are used for the computation of all the efficiencies accounted for in the model. The topography of the site is accounted for by defining the elevation, z, given x and y. Thus,

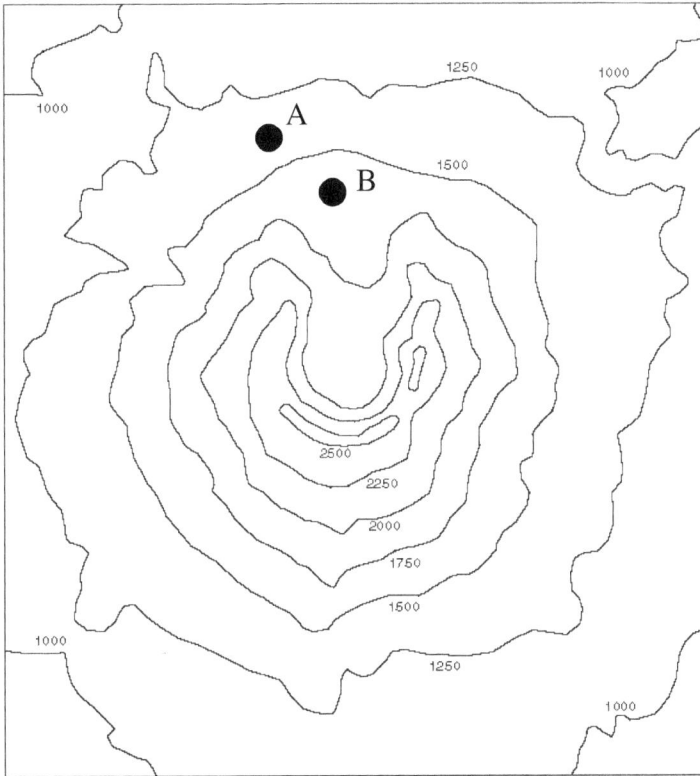

Figure 3.11: Heliostat field topography consideration

for a flat site, $z = 0$ for all x and y. As an illustration of this, consider Figure 3.11. Suppose that this contour diagram represents a site where heliosats are to be placed. The numbers along each contour line represent the height of the physical site above sea level. Each x and y co-ordinate on the site has an associated height above sea level given by the contour lines. Point A, for example, whatever its co-ordinates are, will have a height of around 1300 m above sea level.

Suppose that during optimization, a heliostat moves from point A to point B, the x and y co-ordinates of the heliostat will change because of the optimization procedure. However, the height of the heliostat, or z-value, is dependent on the topography of the site given by the contour diagram of Figure 3.11. Thus the z-value is a function of the x and y values. Table 3.1 indicates expressions that can be used for topography considerations. More complex topography can be accounted for by expressing the site elevation, z, in terms of x and y or by writing lookup tables that represent the contour diagrams of a site.

During optimization, the heliostat co-ordinates, x and y, are variables subject to change as the optimization proceeds. The elevation, z, of the heliostat is redefined by a topography expression (such as those in Table 3.1) after each iteration, given the

Table 3.1: Site topography expressions

Site Topography	Elevation Expression
Slope down 1m/100m East to West	$z = -\frac{x}{100}$
Slope down 1m/100m North to South	$z = -\frac{y}{100}$

new x and y co-ordinates. Thus, topography can be accounted for in the optimization process.

3.11 Flux limit

The amount of energy that may be delivered to the receiver is a function of the material properties of the receiver. In practice, the flux limit is never exceeded in order to prevent thermal failure of the receiver material. This is achieved by defocussing some of the heliostats at times when intercepted radiation might approach the flux limit [46].

Table 3.2 shows typical values for the flux limit. The table also indicates that this value has been increasing over the years. Provision to include such a limit was added to the model. A limit may be set by simply setting the maximum amount of energy that can be collected at any hour to the flux limit of the material being used.

Table 3.2: Incident flux limitations on central receivers [46]

Project	Solar Two	Solar Tres	Solar 50	Solar 100
In service date	1996	2004	2006	2008
Receiver peak incident flux [MW/m^2]	0.8	0.95	1.2	1.4

3.12 Computer Code

The model was programmed for processing on a computer using Octave and Fortran. The sun vector model was programmed in Octave. The Octave script calculates the sun position in vector format at each hour of the year. The output is a text file containing all the data. This data is then used in the Fortran code, which calculates the annual intercepted energy.

To determine the sun vector at each hour, the Octave code requires the following input parameters:

1. The longitude on which the time zone is based at the site, L_{st}

2. The longitude of the site, L_{loc}

3. The latitude angle, ϕ

The Fortran code uses the sun vector information from the Octave code, along with the solar resource information for the site, to calculate the annual intercepted energy of the system using the model presented in this chapter. The Fortran code requires the following input parameters:

1. Mirror type (flat or canted)

2. Field direction (Southern or Northern hemisphere)

3. Tower height

4. Heliostat height

5. Heliostat width

6. Number of heliostats

7. Number of hours for calculation

8. Receiver width or height

9. Receiver x co-ordinate

10. Receiver y co-ordinate

11. Receiver type (external cylindrical or flat)

The Fortran code uses these inputs and calculates the annual intercepted energy following the steps of the model presented in this chapter. Both codes, Octave and Fortran, are included in Appendix B. A sample calculation is also included.

3.13   Summary

The intercepted energy from each heliostat can be calculated by geometric analysis, which can be done at each hour. The results of these hourly calculations can be summed up to give an annual total. All the necessary effects are taken into account without much difficulty. As predicted by the literature, blocking and shading calculations are the most lengthy.

Computer scripts have been written for computation of each of the effects. Various fields can be modeled and analyzed by appropriate inputs to the scripts.

In the next chapter, the accuracy of the model is examined by comparison with ray tracing.

CHAPTER 4

MODEL VALIDATION

To determine the validity of the receiver plant model, the model was compared with results of simulations done by three validated ray tracing software programs. One was developed by Bode and Gauché [17], referred to herein as "Bode's ray tracer". A second, called "Tonatiuh", was developed by Blanco *et al.* [43] and a third "SolTrace" [47], developed by the National Renewable Energy Laboratory (NREL) based in the USA. This chapter describes the simulations performed and the results obtained.

4.1 Single-heliostat Test Cases

A simulation of heliostats laid out in a field was done as depicted in Figure 4.1. The interception efficiency (the product of the individual optical efficiency values) for each heliostat was determined using the Fortran model for four different hours of the year in Upington. Further details of the test are summarized in Table 4.1.

The interception efficiencies were compared with interception efficiencies obtained by simulating the same field using Bode's ray tracer. The results are shown in Figure 4.2. The analysis model correlates closely with Bode's ray tracer. The mean

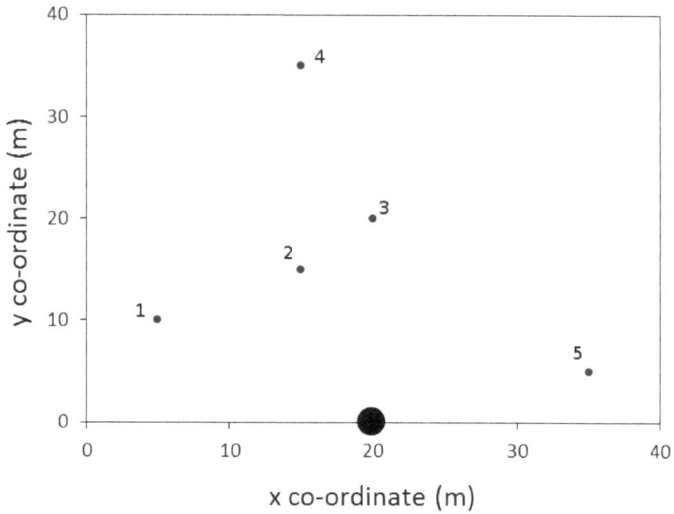

Figure 4.1: Layout of single heliostat test case

Table 4.1: Single-heliostat test case specifications

Location		Heliostats		Receiver	
Latitude	$28°26'$S	Count	5	Tower Height	15m
Longitude	$21°15'$E	Height	1m	Type	External cylindrical
Site width	40m	Width	1m	Diameter	3m
Site length	40m	Geometry	Flat	Height	3m

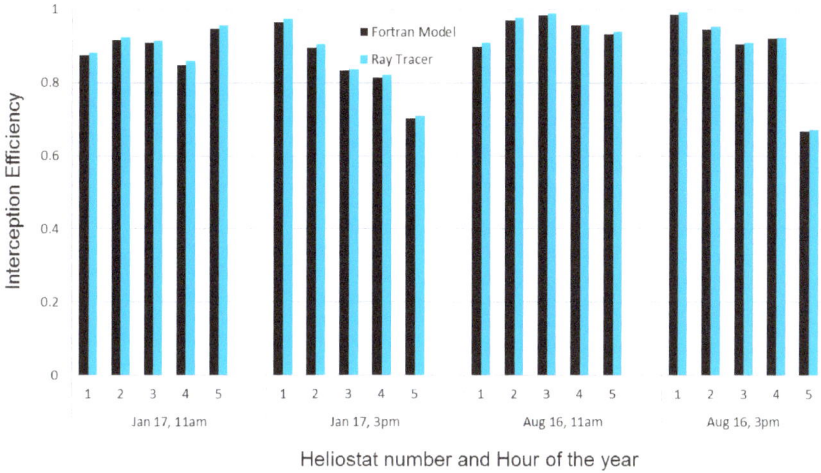

Figure 4.2: Interception efficiency of single heliostat test cases

error of the 20 cases is a 0.79% under-prediction with a standard deviation of 0.3%. This correlation between the two data sets indicates the validity of the solar resource model, which calculates the sun position, and the cosine efficiency model.

4.2   Two-heliostat Test Cases

The interaction between two heliostats causes blocking and shading. Two heliostats placed close enough to each other to cause blocking were simulated to determine the intecepted energy over the course of a spring day in Upington. Figure 4.3 shows the comparison in efficiencies predicted by the Fortran model and SolTrace. The mean error between the Fortran model and SolTrace for these 12 data points was determined to be 0.45% with a standard deviation of 3.1%.

To determine the accuracy of the model on a system level with the effects of blocking included, the intercepted energy from the two-heliostat field onto a flat receiver was determined and compared with SolTrace. The receiver was specified to be large enough to deactivate the effects of spillage. DNI at each hour was taken as $1000\,\mathrm{W/m}^2$. Figure 4.4 shows the comparison between the Fortran model and SolTrace in predicting the intercepted energy at each hour of the day. The figure also gives the amount of power that was blocked by the heliostat causing the blocking. The blocked power was calculated in SolTrace.

Over the 12 hour day, the mean error between the Fortran model and the SolTrace model was 0.3% with a standard deviation of 1.0%.

These results also give an indication of the effectiveness of the shading efficiency calculation. Blocking and shading calculations are similar. The only difference is the

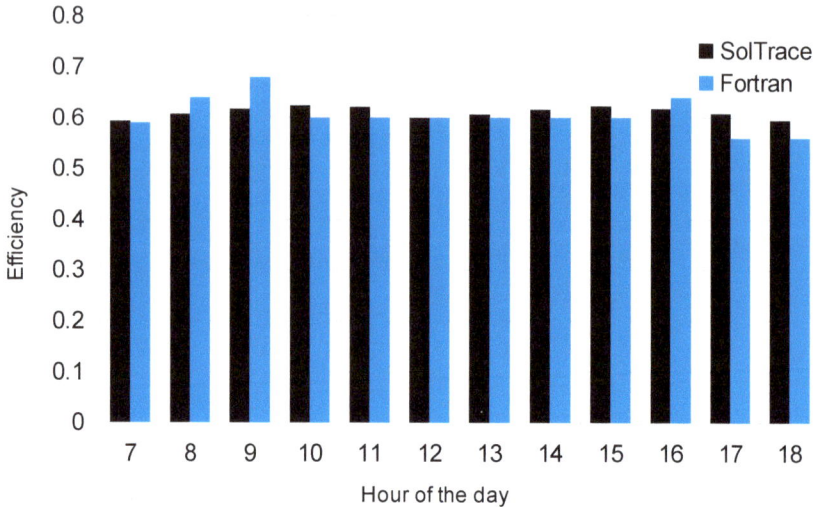

Figure 4.3: Comparison of blocking efficiencies predicted by Fortran model and SolTrace

vector that is used to construct the interception lines from the heliostat potentially being blocked or shaded.

4.3   Spillage Test Cases

To determine the validity of the spillage model, a single heliostat placed at varying distances from the receiver was simulated. The spillage efficiency of the heliostat was determined using the Fortran model and then using the ray tracing software Tonatiuh [43]. The comparison of the two data sets is depicted in Figure 4.5.

The mean difference between the model and Tonatiuh is 2.1% with a standard deviation of 6.7%. The model is less accurate at distances closer to the target. To understand the reason behind this, photon maps of the receiver were constructed in Tonatiuh. These images are shown in Figure 4.6. The images show the difference in images cast by heliostats close to the receiver and far away from the receiver.

The thin, elongated image caused by the close heliostat indicated in Figure 4.6b, is due to the elliptical shape caused by the angle that the image makes with the receiver. The farther heliostat, Figure 4.6a, causes an image that is more spread-out, and the image is spilled over all sides of the receiver.

These results indicate that the spillage model used is more accurate at predicting the spillage efficiency of heliostats farther away. The model tends to "penalize" heliostats that are close to the receiver tower. In optimization, this would mean that the model will tend to move the heliostat away from the receiver tower if they are close. The model, however, is a reasonable approximation of the effect of spillage.

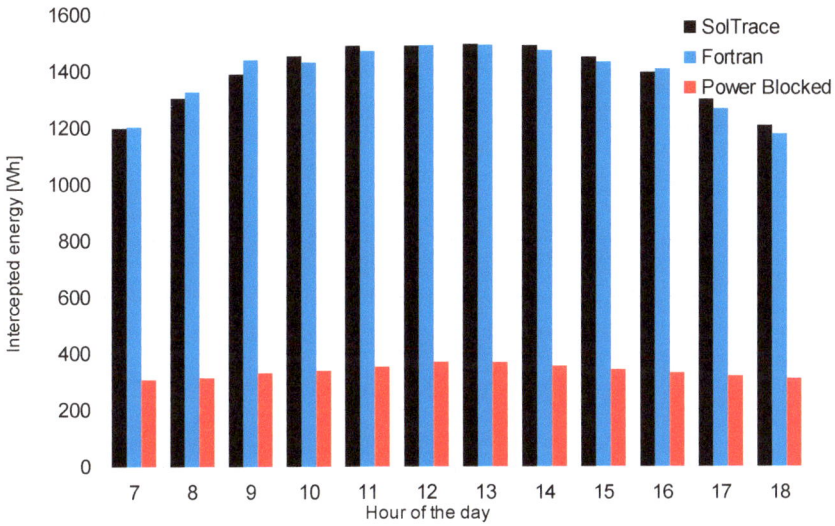

Figure 4.4: Comparison of Fortran model with SolTrace with blocking effects included

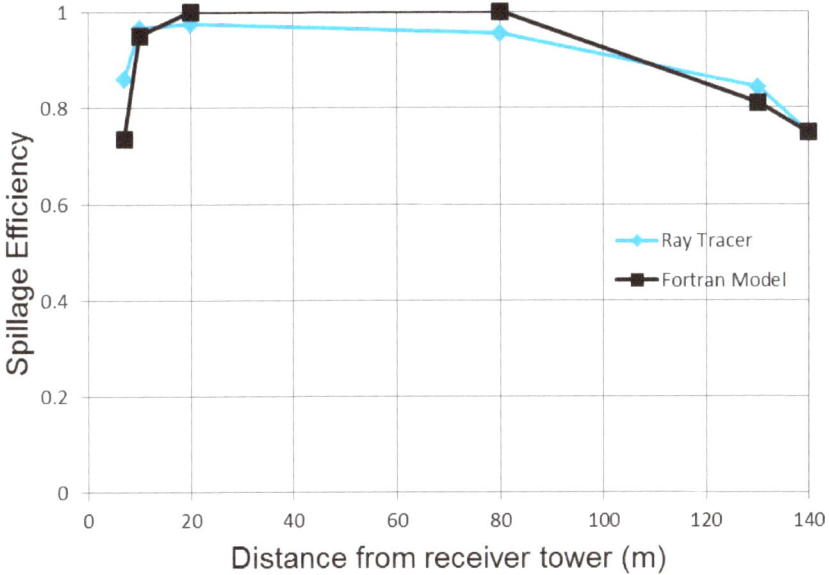

Figure 4.5: Spillage efficiencies from Tonatiuh and Fortran model

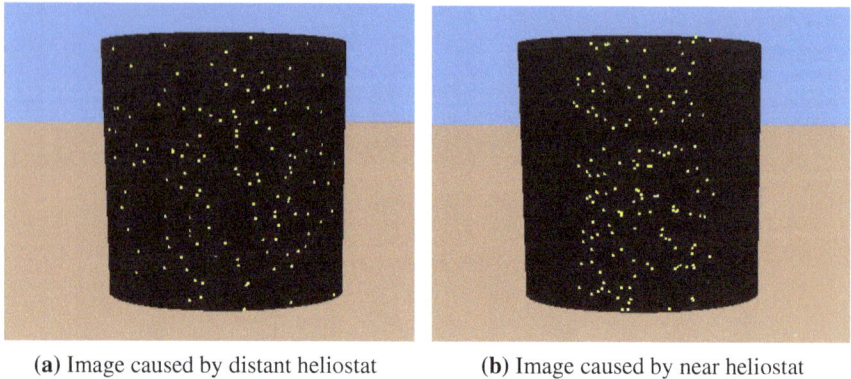

(**a**) Image caused by distant heliostat   (**b**) Image caused by near heliostat

Figure 4.6: Spillage at different distances

4.4 System Validation

To determine the validity of the model on a system level, a hypothetical plant was created and evaluated with the Fortran model and with the ray tracing software SolTrace. The plant consisted of 100 randomly located heliostats. The field is shown in Figure 4.7. The performance of the plant was evaluated over the course of a day. The comparison with the ray tracer is indicated Figure 4.8. Further details about this simulation are included in Table 4.2.

Table 4.2: System validation field specifications

Location		Heliostats		Receiver	
Latitude	28°26′S	Count	100	Tower Height	50m
Longitude	21°15′E	Height	2m	Type	Flat
Site width	50m	Width	2m	Width	1m
Site length	50m	Geometry	Flat	Height	1m

The average difference between the SolTrace model and the Fortran model was 0.4% with a standard deviation of 3.3%. The intercepted energy collected by this field over the course of the day was calculated to be 148 kWh with SolTrace, and 147 kWh with the Fortran model. The Fortran model under-predicted the intercepted energy, relative to SolTrace, by 0.4%. A similar accuracy would be expected for the annual intercepted energy.

During ray tracing the number of rays used was 100 000. To get a good representation of the system, either a large number of rays should be used or the mean of a sufficiently large population of ray traces should be determined. Making use of the latter method, 30 traces were done for a chosen hour and a statistical analysis of the data was performed. From this analysis it was found that there was a 95% confi-

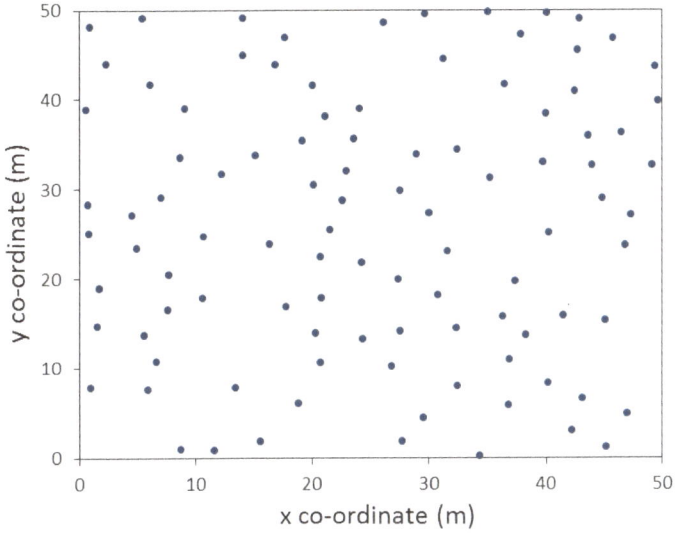

Figure 4.7: Random field used for system validation

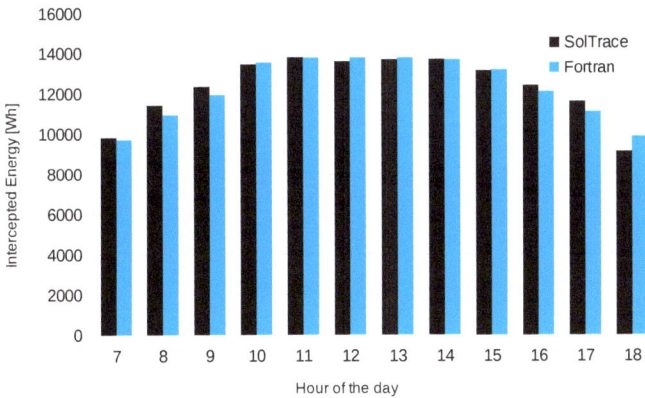

Figure 4.8: Comparison of Fortran and SolTrace system performance evaluation

dence level that any value from the ray tracer would be within 0.4% of the mean of a population of ray traces.

4.5 Summary

The receiver plant model correlates well with ray tracing in its calculation of the sun position and cosine efficiency, as well as with blocking calculations. Spillage efficiency calculations are less accurate. For heliostats that are far from the receiver, the spillage calculation is more accurate and less so for heliostats that are close to the receiver. On a system level the accuracy level is also sufficient, and the model shows the necessary trends needed for the purpose of this approach. The model can thus be used for optimization.

In the following chapter, the model is combined with an optimization algorithm and optimization is conducted using the model. Ray tracing is a far better analysis tool and can be used to determine the actual intercepted energy of a heliostat field. This model can be used as a driver for the optimization and the results can be analyzed with a ray tracer to determine the final intercepted energy.

CHAPTER 5

OPTIMIZATION

This chapter describes how the receiver plant model is cast as an optimization problem and how the optimization can be conducted using an optimization algorithm that is suitable to the form of the optimization problem. For the optimization conducted, an optimization algorithm developed by Groenwold and Etman [48], denoted SAO*i*, is used. The description of the algorithm given herein is adapted from this source.

5.1   The Optimization Problem

To formulate the plant model into an optimization problem, the model is written as a function of its variables. As presented in Chapter 3, the intercepted energy is calculated as follows:

$$I = A \sum_{k=1}^{192} \text{DNI}_k \left(\sum_{i=1}^{m} \eta_{c_{i,k}} \eta_{a_i} \eta_{sp_i} \eta_{b_{i,k}} \eta_{s_{i,k}} \right) \tag{5.1}$$

Note that the number of hours has been reduced from 8760 to 192. This has been done to reduce the computational expense of the problem. Duffie and Beckman [42] suggest that 12 typical days of the year (1 day per month) can be used when doing an annual analysis. The dates suggested by the author have been employed. Furthermore, only the 16 daylight hours of each day are considered. Solar irradiation occurs only during the daytime, so the nighttime would have no influence on the total intercepted energy of the plant. Including the night hours would only result in an unnecessarily higher computational overhead.

Since each efficiency contained in the model is dependent on the location of each heliostat within the field relative to the tower, the annual intercepted energy is a function of the co-ordinates of the heliostats. The expressions vary depending on other plant characteristics such as the receiver type and heliostat shape. Stated as a function of the design variables, the annual intercepted energy may be expressed as follows:

$$f(x) = -I = -A \sum_{k=1}^{192} \text{DNI}_k \left(\sum_{i=1}^{m} \eta_{c_{i,k}} \eta_{a_i} \eta_{sp_i} \eta_{b_{i,k}} \eta_{s_{i,k}} \right) \qquad (5.2)$$

The reason for the negatives will be clarified in a subsequent section. To evaluate the annual intercepted energy of a field, the function requires as input the vector x. This vector contains the x and y values of the heliostats' positions within the field. Thus, for a field with m heliostats, x has $n = 2m$ dimensions. Other plant characteristics such as tower height, receiver size or aperture area, and receiver inclination, to name a few, may also be added to the input vector as design variables. Each added plant characteristic will increase the variable count by 1. Thus, with, for example, 3 more design variables, x will have $2m + 3$ dimensions. In the present model, these other plant characteristics have not been included; only the heliostat locations have been used as design variables.

For a pattern method of optimization, the vector x is itself a function of the parameters that define the pattern [34] and is, therefore, dependent on them. During optimization, from iteration to iteration, only the pattern variables are altered. Altered pattern parameters produce different x and y values in the vector x. Intercepted energy then, in effect, becomes a function of the pattern parameters. The objective then of the optimization is to determine the optimum values for these pattern parameters. Optimum pattern parameters define an optimum input vector x, which then is the optimal adaptation of the pattern for the problem. The resulting vector x delivers the best value for the function that can be obtained by that pattern. As Buck [34] has pointed out, and as will be shown here, this resulting field may not necessarily be optimal in terms of individual heliostat locations.

For the optimization method used herein, a free or non-restricted method, similar to what is presented by Buck [34], is applied. That is, the vector x is independent. There is one key difference though. Whereas Buck allows freedom within a small area surrounding the heliostat, the free variable method allows for complete freedom of the variables; variables may take on any value within the site boundary. This is done so that optimal values for individual heliostat locations may be obtained.

The freedom of variables has an effect on the computational complexity of the problem. Each of the x and y values contained in the vector are considered design variables and, thus, may be varied independently from iteration to iteration to determine the optimal location of each heliostat within the field for maximum annual intercepted energy. This means that the number of design variables in the optimization is extremely large compared to, for instance, the pattern method.

Evaluation of the function in equation 5.2 from iteration to iteration can be done by constructing a computer script which evaluates each of the efficiencies given the vector x. The current author constructed such a script in Fortran.

5.2 Constraints

Essentially, there are two constraints that need to be considered in the optimization. Stated in plain terms, the two constraints are as follows:

- Each heliostat must be a certain minimum distance from the next heliostat to prevent collision between the heliostat surfaces during operation

- Each heliostat must be a certain minimum distance from the central receiver to prevent collision between the heliostats and the receiver tower

Though these constraints are two in essence, both need to be imposed onto every heliostat. Thus, for m heliostats, the actual number of constraints is, for the first constraint

$$(m^2 - m)/2 \tag{5.3}$$

and for the second

$$m \tag{5.4}$$

giving a total of

$$(m^2 - m)/2 + m \tag{5.5}$$

The subtraction by m in equation 5.3 is to exclude the heliostat being evaluated. That is, a heliostat must be a certain minimum distance from every other heliostat, itself excluded. The division by 2 is due to the fact that only one permutation of this constraint is needed. In other words, the distance from heliostat 1 to heliostat 2 is the same as the distance from heliostat 2 to heliostat 1.

The minimum distance that each heliostat must be from every other heliostat as well as from the tower can be taken as the length of the diagonal of the heliostat surface. This will ensure that heliostats do not interfere with each other during operation. Thus, the constraints can be expressed as follows, for the first constraint

$$||(x_i, y_i) - (x_j, y_j)|| \geq a$$

or

$$-||(x_i, y_i) - (x_j, y_j)|| + a \leq 0 \tag{5.6}$$

$$i = 1, 2, \ldots, m-1 \text{ and } j = i+1, i+2, \ldots, m$$

and for the second constraint

$$||(x_i, y_i) - (x_R, y_R)|| \geq b$$

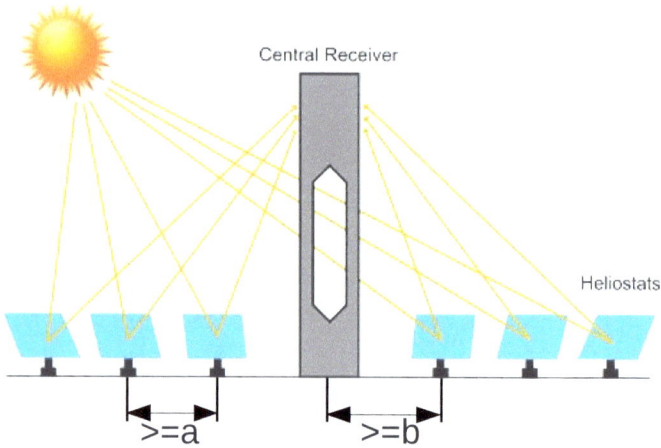

Figure 5.1: Constraints [49]

or

$$-||(x_i, y_i) - (x_R, y_R)|| + b \leq 0 \qquad (5.7)$$

$$i = 1, 2, \ldots, m$$

where the subscript R refers to the receiver. The constant a is taken to be the length of the diagonal of a heliostat and b is half the diagonal plus half the width of the receiver tower. The constraints are depicted in Figure 5.1.

It was found necessary to include all the constraints represented in equation 5.6. Heliostat constraints, for example, cannot be split into sections of the land area else heliostats on opposite ends of the field might gravitate towards each other during optimization. This will be discussed in a subsequent section.

The result is a large number of constraints. Pattern methods have these constraints implicit in their definition. Thus, in the optimization of a pattern, these constraints are redundant. Other constraints may also be added to these, for example, to make provision for roads. Each additional requirement will raise the constraint count by m.

It is vital to select an appropriate optimization method for the problem due to the large number of variables and constraints.

5.3 Optimization Method

To perform optimization the receiver plant model is cast as an inequality constrained nonlinear optimization problem, P_{NLP} of the following form:

$$\begin{aligned}
\min \quad & f_0(\boldsymbol{x}) \\
\text{subject to} \quad & f_j(\boldsymbol{x}) \le 0 \qquad j = 1, 2, \ldots, w \\
& \check{x}_i \le x_i \le \hat{x}_i \qquad i = 1, 2, \ldots, n
\end{aligned} \qquad (5.8)$$

where $f_0(\boldsymbol{x})$ is a real valued scalar objective function, and $f_j(\boldsymbol{x})$, $j = 1, 2, \ldots, w$ are w inequality constraint functions. $f_0(\boldsymbol{x})$ and $f_j(\boldsymbol{x})$ depend on the n real (design) variables $\boldsymbol{x} = \{x_1, x_2, \ldots, x_n\}^T \in \mathcal{X} \subset \mathcal{R}^n$, which define the $\check{x}_i$ and $\hat{x}_i$ respectively as the lower and upper bounds on variable x_i. In a typical field optimization problem, the lower and upper bounds would represent the site boundaries.

Note that in the problem, $x_1, x_2, \ldots, x_m$ are the x values of the positions of each of the heliostats and $x_{m+1}, x_{m+2}, \ldots, x_n$ are the y values. Also, the objective is to minimize the function, hence the negatives in equation 5.2. By minimizing the function, $f(\boldsymbol{x})$, in the optimization, the annual intercepted energy is maximized.

The constraint functions, $f_j(\boldsymbol{x})$, $j = 0, 1, 2, \ldots, w$, are known to be differentiable using either finite differences or, preferably, automatic differentiation. It is also known that f_0 and f_j require a fairly expensive computational simulation. In addition, it is known that the number of design variables and constraints are high, which effectively disqualifies zeroth order methods.

Groenwold et al. [30] have proposed to use diagonal quadratic approximations to approximations based on arbitrary (albeit separable) intervening variables, rather than an approximation based on a specific intermediate variable. These formulations have the advantage that a single dual statement may be used to capture the essence of the arbitrary intermediate variables considered. Only diagonal Hessian information is required. In addition, since the approximations are (diagonal) quadratic, they may easily be transformed into quadratic programs (QPs). This makes it possible for them to be treated by high-quality existing solvers [50; 51].

5.4 Algorithm SAO*i*

Algorithm SAO*i* makes use of a sequential approximate optimization as a solution strategy for problem 5.8. Instead of using diagonal quadratic function approximations, SAO*i* constructs successive approximate subproblems $P[k]$, $k = 1, 2, 3, \ldots$ at successive iteration points, $\boldsymbol{x}^{\{k\}}$, that are inexpensive to evaluate. The solution of subproblem $P[k]$ is denoted $\boldsymbol{x}^{\{k*\}} \in \mathcal{R}^n$, and is to be obtained using a suitable continuous programming method. The minimizer of subproblem $P[k]$ is $\boldsymbol{x}^{\{k*\}}$, which is then ready to become the starting point $\boldsymbol{x}^{\{k+1\}}$ for the subsequent approximate subproblem $P[k+1]$. The approximations and subproblems considered in algorithm SAO*i* are summarized in the following subsections.

5.4.1 Diagonal Quadratic Approximations

SAO*i* constructs diagonal approximations, $\tilde{f}(x)$, to the objective function, $f_0(x)$, and all the constraint functions $f_j(x)$ as

$$\tilde{f}_j(x) = f_j^{\{k\}} + \sum_{i=1}^{n}\left(\frac{\partial f_j}{\partial x_i}\right)^{\{k\}}(x_i - x_i^{\{k\}}) + \frac{1}{2}\sum_{i=1}^{n} c_{2i_j}^{\{k\}}(x_i - x_i^{\{k\}})^2 \quad (5.9)$$

with $f_j^{\{k\}} = f_j(x^{\{k\}})$ and $c_{2i_j}^{\{k\}}$ the approximate second order diagonal Hessian terms or curvatures.

To ensure strict convexity of each and every subproblem P[k] to be considered, it is necessary to invariably enforce

$$\begin{aligned} c_{2i_0}^{\{k\}} &= \max(\epsilon_0 > 0, c_{2i_0}^{\{k\}}) \\ c_{2i_j}^{\{k\}} &= \max(\epsilon_j \geq 0, c_{2i_j}^{\{k\}}) \quad j = 1, 2, \ldots, m \end{aligned} \quad (5.10)$$

with the ϵ_j, $j = 0, 1, 2, \ldots, m$ prescribed and "small" (or zero). In other words, the approximate objective function $\tilde{f}_0$ is strictly convex, while the approximate constraint functions $\tilde{f}_j$, $j = 1, 2, \ldots, m$ are convex or strictly convex.

5.4.2 Estimating the Higher Order Curvatures

Key to algorithm SAO*i* is to obtain approximate higher order curvatures $c_{2i_j}^{\{k\}}$ without the user providing, or the algorithm storing, explicit second order information. There are many possibilities for doing this, including the use of finite differences for estimating the diagonal Hessian terms only, and quasi-Cauchy-updates [52], to name a few.

The simplest possible strategy, arguably, is to construct a spherical quadratic approximation by selecting $c_{2i_j}^{\{k\}} \equiv c_{2_j}^{\{k\}}$ for all i. This requires the determination of the single unknown $c_{2_j}^{\{k\}}$, easily obtained by, for example, enforcing this condition:

$$\tilde{f}_j(x^{\{k-1\}}) = f_j(x^{\{k-1\}}). \quad (5.11)$$

Groenwold *et al.* [30] provide more details. A detailed description of the steps of the algorithm is given in Appendix C.

5.5 Fortran Code

The receiver plant model code was combined into the SAO*i* code for optimization problems. SAO*i* was written in Fortran. For this reason it was simple to combine the receiver plant model into the optimization code.

5.6   Results

A number of optimization problems were constructed and carried out for evaluation of the optimization method. This section describes each of these attempts and the results obtained. The specifications of the machines used for these optimization runs are given in Table 5.1. Parallelization was not utilized.

Table 5.1: Machine specifications

Machine	CPUs	Memory	Operating System
Toshiba Laptop	Intel Core i5 M560 2.67 GHz	4 GB	Ubuntu Linux 12.10, 13.04, 13.10 (64-bit)
Dell Desktop	Intel Xeon 3.73 GHz	32 GB	openSuse Linux 11.4 (64-bit)
HPC	Intel Xeon E5440 2.83 GHz	16 GB	

5.6.1   Hypothetical Plant

A small hypothetical plant was modeled as a test case for optimization. Site information for Upington, South Africa, was used. The plant aperture area was taken to be $400 \, \text{m}^2$ with a site length and width of 40 m each. Further specifications of the plant are tabulated in Table 5.2.

Table 5.2: Hypothetical plant test case specifications

Location		Heliostats		Receiver	
Latitude	$28°26'$S	Count	400	Tower Height	15m
Longitude	$21°15'$E	Height	1m	Type	External cylindrical
Site width	40m	Width	1m	Diameter	3m
Site length	40m	Geometry	Flat	Height	3m

To start the optimization, a random field was created. This field is shown in Figure 5.2a. The optimization converged after 100 iterations and approximately 39 hours. The optimization produced the field shown in Figure 5.2b. The optimized field is a 21% improvement over the random field in terms of annual intercepted energy. Upon close inspection of Figure 5.2b, a pattern can be seen resembling the layout of sunflower petals.

This result indicates, firstly, that the free variable method is possible. Secondly, as a result of the pattern that is evident, it indicates that there is rationality in using a pattern for a heliostat field; elegant patterns, that perhaps resemble natural patterns, are more optically efficient than random fields.

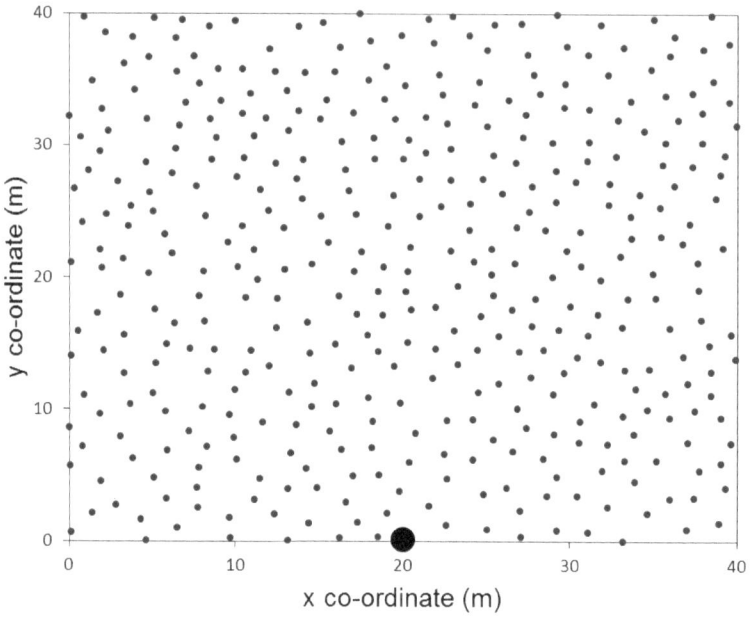

(a) Random field

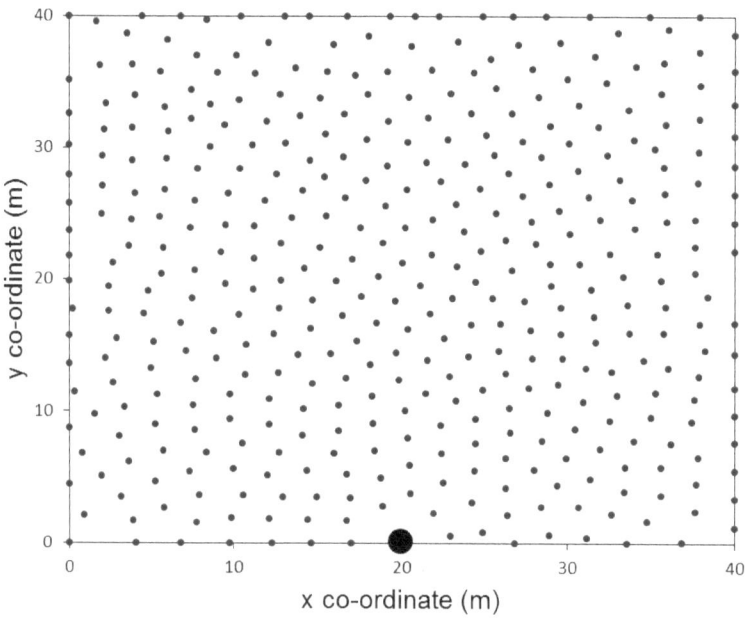

(b) Optimized field

Figure 5.2: Hypothetical plant optimization

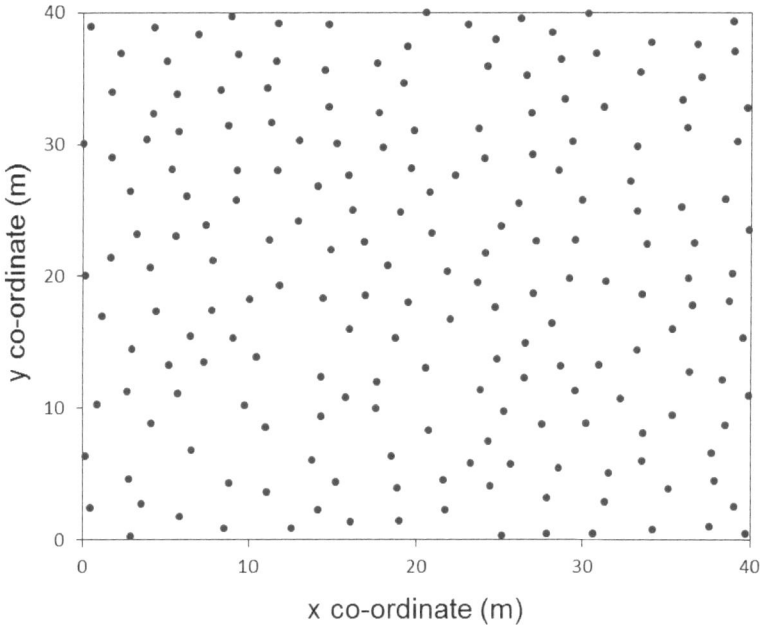

Figure 5.3: Random field

5.6.2   Receiver Type Effects

To evaluate the effect that the receiver type has on the heliostat field layout, a similar plant was modeled using different receiver types. Two cases were modeled. In one case an external cylindrical receiver was assumed, and in the other case a flat receiver was assumed. Details of the plant are indicated in Table 5.3.

Table 5.3: Receiver type effects case specifications

Location		Heliostats		Cylindrical receiver		Flat receiver	
Latitude	28°26′S	Count	200	Diameter	3m	Width	3m
Longitude	21°15′E	Height	2m	Height	3m	Height	3m
Site width	40m	Width	2m				
Site length	40m	Geometry	Flat	Tower Height		15m	

The initial field used in this optimization was, once again, a random field. This field is shown in Figure 5.3. The two fields produced by the optimization runs are shown in Figure 5.4. The cylindrical receiver caused a more spread-out, surrounding field; the flat receiver less so.

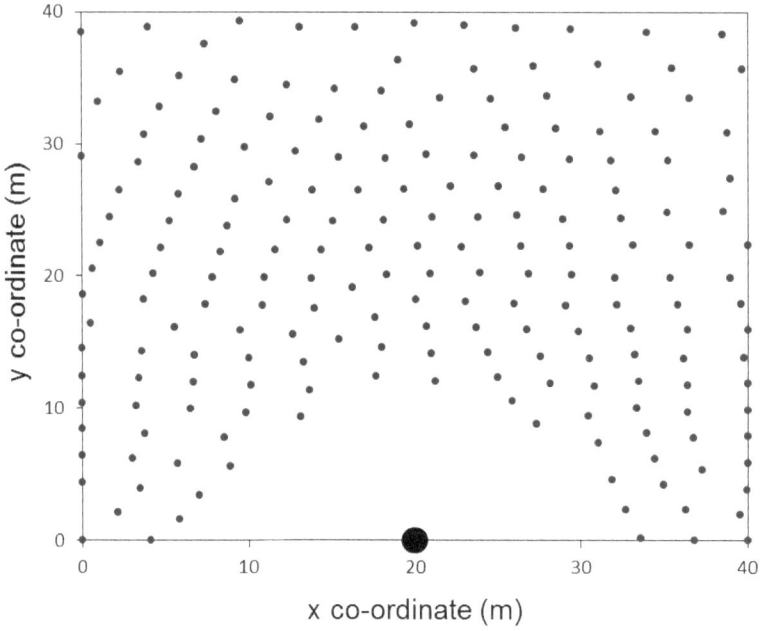

(a) Cylindrical receiver

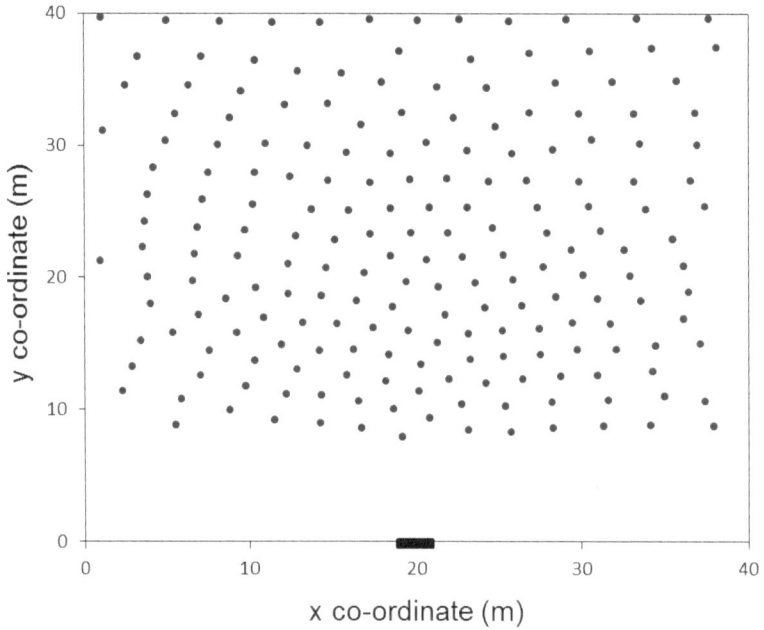

(b) Flat receiver

Figure 5.4: Fields optimized for different receiver types

This result indicates that the free variable method is one that is problem-oriented as it tends to produce a field that is suited to the characteristics of other plant components.

This optimization was done with 200 heliostats—half the number used in the first hypothetical plant presented in the previous section. For this reason, there was more space available for the heliostats. As a result, the heliostats tended to move away from the receiver tower as was expected due to the penalization caused by the spillage model.

5.6.3 Sensible Start

To determine the behavior of the optimization when starting from a sensible field, an optimization was done starting a field with a staggered pattern. This field is shown in Figure 5.5a. The tower is located at (10,0). After optimization, the starting field and the improved field were analysed in SolTrace over the course of a day. The initial field was able to collect 824.6 kWh and the improved field 846.6 kWh, an improvement of 2.67%. The improved field is shown in Figure 5.5b. The SolTrace comparison of the two fields is shown in Figure 5.6.

5.6.4 Constraint Relaxation

One of the time consuming operations of the optimization is the evaluation of the constraints. It was shown earlier that there is a large number of constraints. A test was carried out to determine whether or not the number of constraints could be reduced. The first basic constraint requires that each heliostat be a certain minimum distance from every other heliostat. To reduce the number of constraints, this requirement was relaxed.

By applying this modification, the total number of constraints was reduced from $(m^2 - m)/2 + m$ to m. The initial field was a random field with 100 heliostats. Further field specifications for this case are included in Table 5.4. The random field is shown in Figure 5.7a and the field after optimization is shown in Figure 5.7b. The tower was located at (25,0).

Table 5.4: Constraint relaxation field specifications

Location		Heliostats		Receiver	
Latitude	28°26'S	Count	100	Tower Height	50m
Longitude	21°15'E	Height	1m	Type	Flat
Site width	50m	Width	1m	Width	2m
Site length	50m	Geometry	Flat	Height	3m

The locations of the heliostats in the optimized field are not close enough to interfere with each other. This means that during optimization, the blocking and shading heuristics of the technical model kept the heliostats from moving too close to each

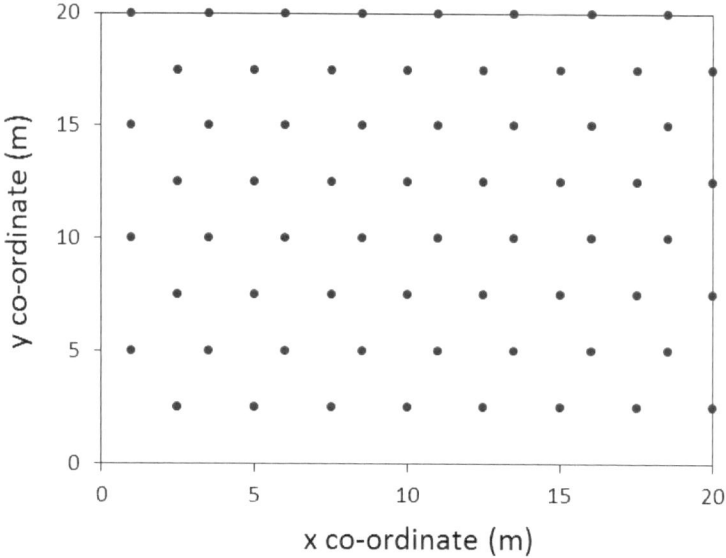

(a) Sensible field

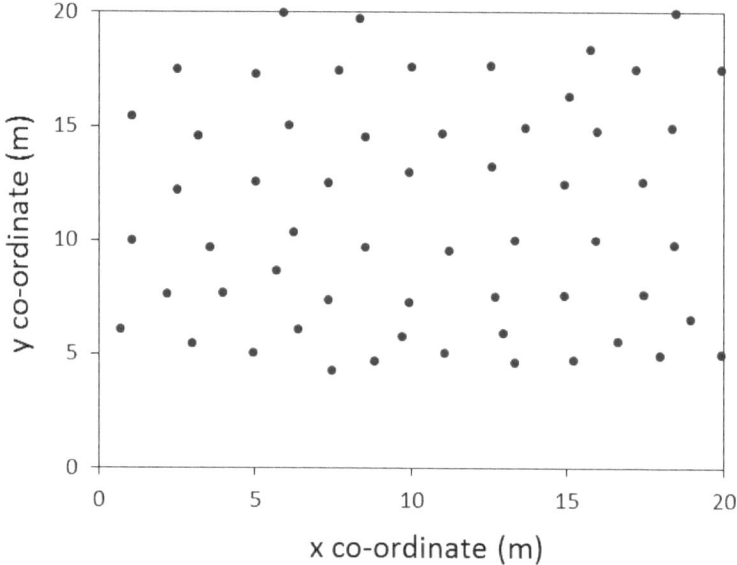

(b) Sensible field after optimization

Figure 5.5: Optimization from a sensible start

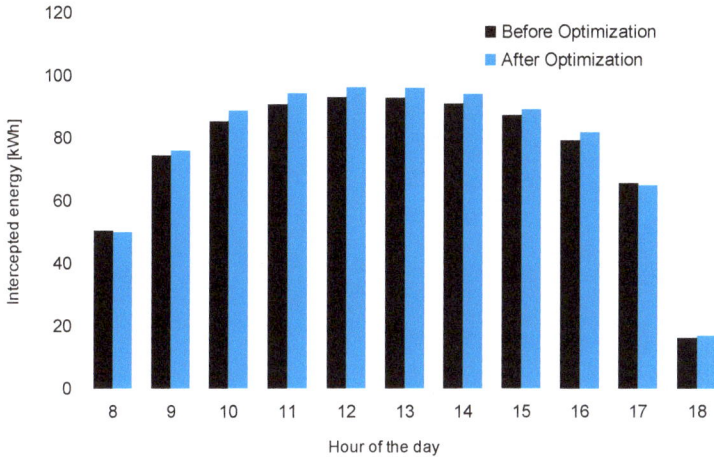

Figure 5.6: Intercepted energy computed in SolTrace before and after optimization of a sensible field

other. The exercise reveals that it may not be necessary to enforce all the constraints. This, however, needs to be tested to a greater degree. Losses from the cosine effect are usually greater than losses from blocking and shading [5]. There may be cases where heliostats move to an area with low cosine loss and trade off blocking and shading losses for low cosine losses which could then lead to interference.

For comparison, the same field was optimized with all the constraints in place. The resultant field from this optimization is indicated in Figure 5.8. This field is different to the field optimized without the constraints, yet both final fields have similar performance levels. Table 5.5 compares various parameters of the two optimization runs. The field optimized without the constraints performs slightly better than the field optimized with the constraints. However, given the accuracy of the model and that the initial field had an annual intercepted energy value of 63.5 MWh, this difference is insignificant.

As expected, the case with the relaxed constraints required less time to reach an optimum. Because of the large amount of time required per iteration, the fully constrained case required 20% more time than the relaxed case albeit with less iterations.

5.6.5 Optimization Validation

The accuracy of the optimization was evaluated by evaluating a field before and after optimization and comparing the results with the ray tracing software SolTrace. The field constructed for evaluation had an initial random start and consisted of 100 heliostats. This is the same field used for system validation in the previous chapter. The specifications of this field are repeated here in Table 5.6. Images of this field before and after validation are shown in Figure 5.9.

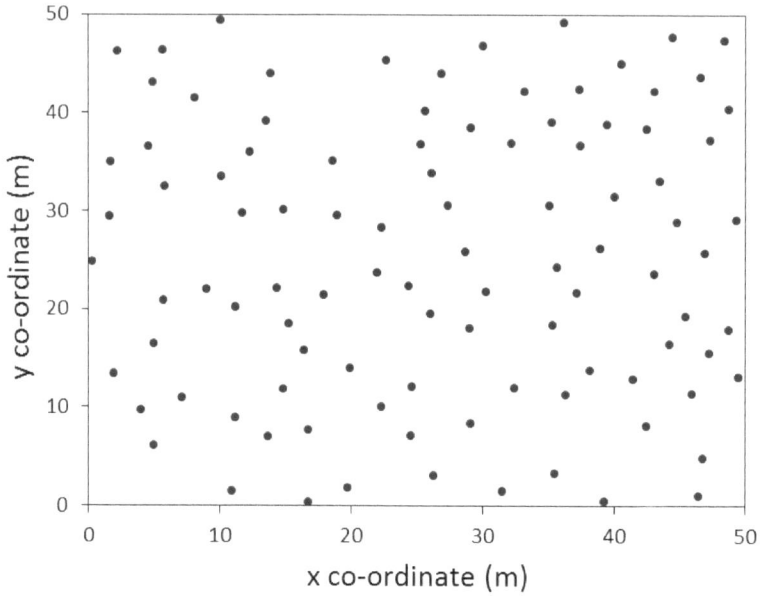

(a) Layout before optimization

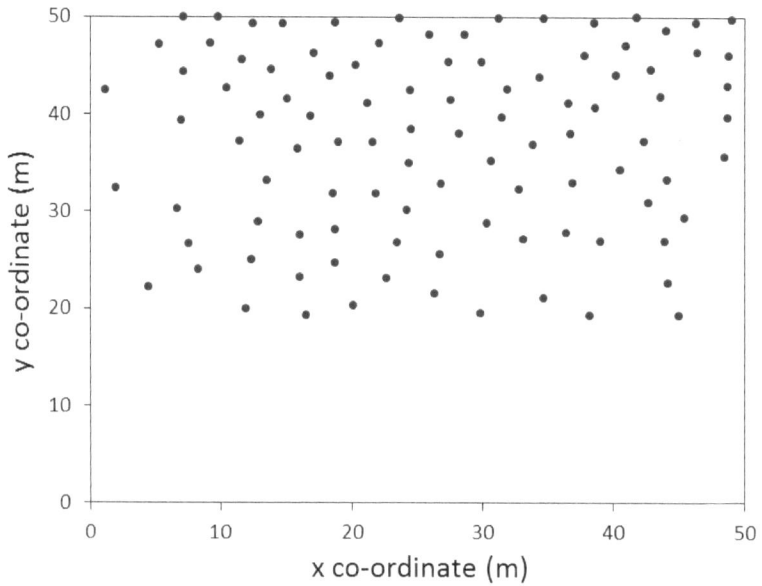

(b) Layout after optimization

Figure 5.7: Field optimization with constraints relaxed

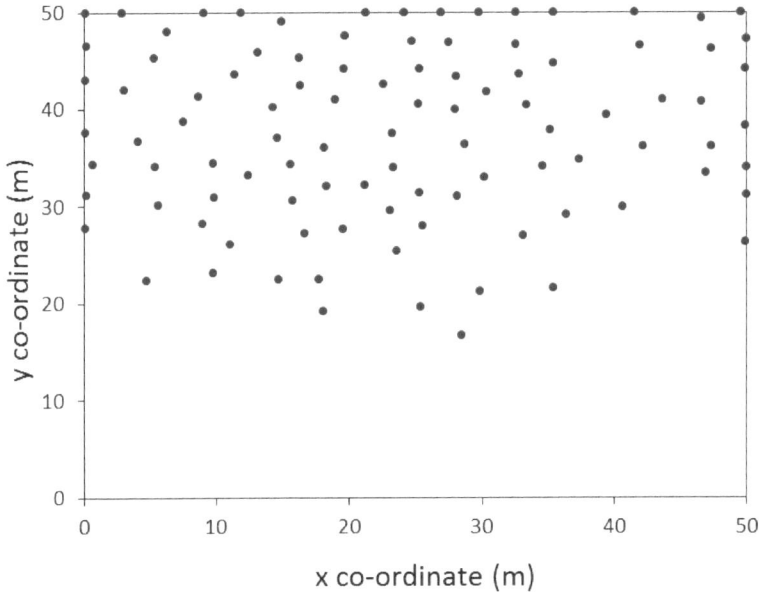

Figure 5.8: Field optimized with all constraints applied

Table 5.5: Comparison of relaxed constraint optimization with fully constrained optimization

Case	Total time	Average time per iteration	Number of iterations	Annual intercepted energy
	[s]	[s]		[MWh]
Constraints applied	17898	365	49	83.9
Constraints relaxed	14109	103	135	82.8

Figure 5.10 shows a comparison of the Fortran model with SolTrace before and after optimization. As mentioned before, the Fortran model under-predicted the intercepted energy, relative to SolTrace, over the course of the day by 0.4%. After optimization, the Fortran model under-predicted once again, relative to SolTrace, by 0.3%. This demonstrates that the optimization combined with the Fortran model does drive the optimization adequately towards a better intercepted energy value.

5.7 Summary

The free variable method is a method of heliostat field layout optimization that follows a more classical optimization approach where heliostats are allowed to move

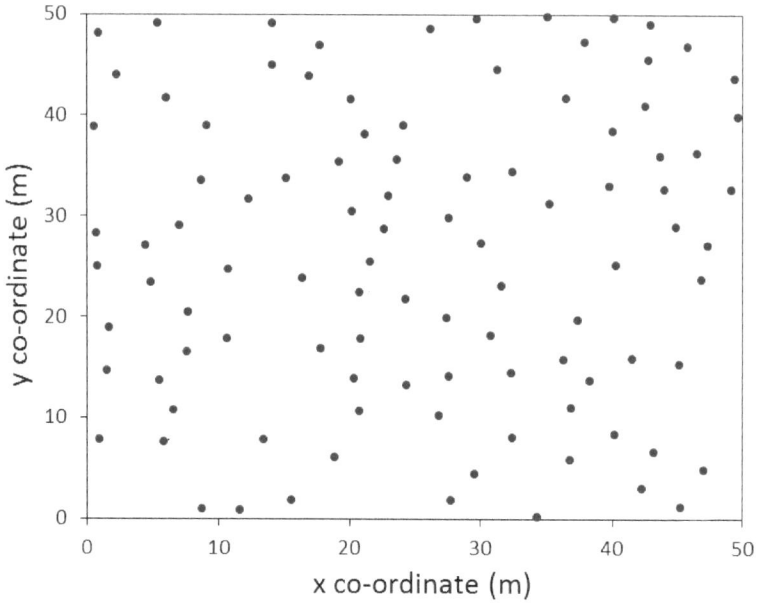

(a) Random field

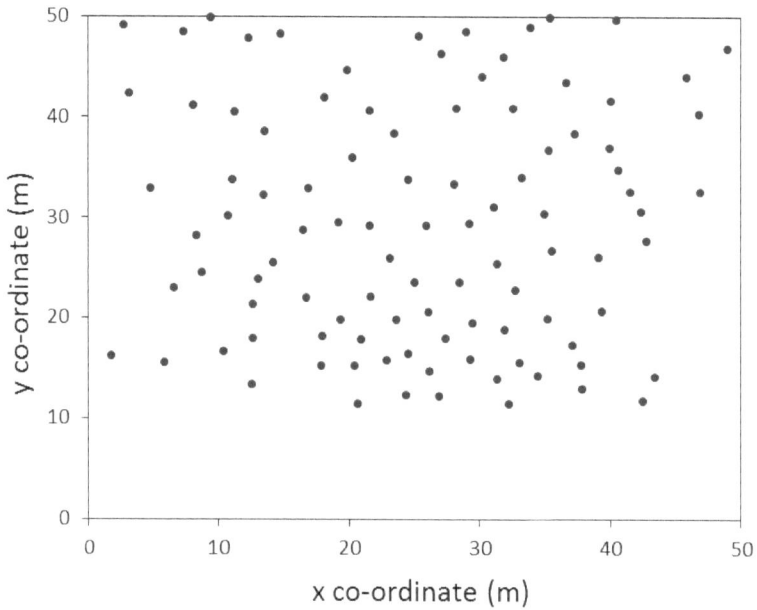

(b) Optimized field

Figure 5.9: Optimization done for validation of optimization

Table 5.6: Optimization validation field specifications

Location		Heliostats		Receiver	
Latitude	28°26′S	Count	100	Tower Height	50m
Longitude	21°15′E	Height	2m	Type	Flat
Site width	50m	Width	2m	Width	1m
Site length	50m	Geometry	Flat	Height	1m

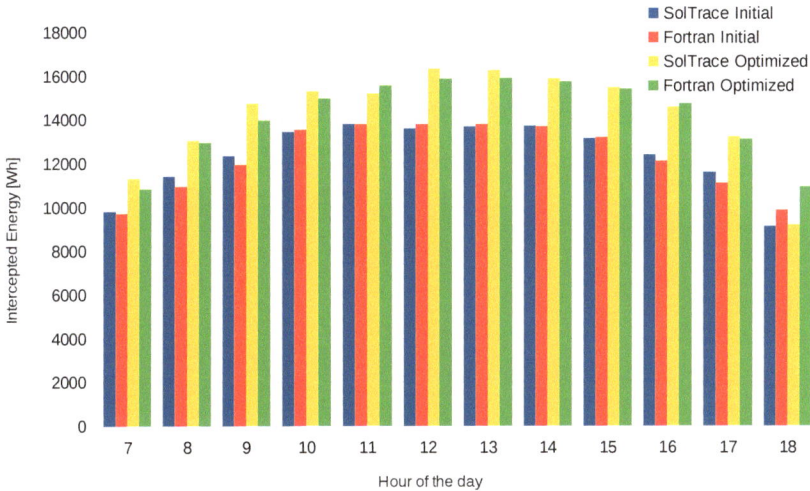

Figure 5.10: Comparison of Fortran model with SolTrace before and after optimization

freely to an optimal location. The method has the disadvantage of having a large number of variables and constraints. It is made possible using a constrained gradient-based optimization algorithm. The SAO*i* algorithm, based on sequential approximate optimization, is one such algorithm.

When starting from a random field, the free variable method tends to produce pattern-like fields that resemble naturally occurring patterns. The method generates fields that are suited to other plant characteristics such as receiver type. The method is time-consuming. This is due to the large number of variables and constraints. The constraints may not be entirely necessary though. The free variable method is also able to improve fields of sensibly arranged heliostats.

In the next chapter, a commercial central receiver plant, PS10, located in Spain, is studied to gauge the strength of the optimization presented in this chapter.

CHAPTER 6

PS10 CASE STUDY

In this chapter, a study of the PS10 field, located in Spain, is presented. The field was analyzed with the plant model presented in Chapter 3 and a redesign was done using the optimization method presented in Chapter 5.

6.1 The PS10 Field

The PS10 concentrating solar power plant is situated in the city of Seville, which is in the south of Spain in Andalusia. The plant is operational and has an electric power rating of 11 MW. The thermal inertia of the plant allows it to run for 50 minutes at 50% load during cloud transients. The plant takes up a land area of approximately 55 hectares and has a total of 624 heliostats, each with an aperture area of $120\,\mathrm{m}^2$ [53]. Figure 6.1 shows an image of the plant.

The original PS10 field was designed using the software tool by Sandia called WinDELSOL1.0 [20]. This code uses a pattern method of optimization [35]. This means that a geometric pattern is assumed and the optimization process determines the best values of the parameters that define the pattern [34]. The optimization provides a field layout that is the best adaptation of the pattern for the plant requirements.

Figure 6.1: The PS10 field [54]

From the literature, such as Buck [34] and Noone *et al.* [20], the PS10 field appears to be the field of choice for benchmarking of optimization techniques. For this reason, the PS10 plant was analyzed using the model and optimization method presented in previous chapters. The specifics of the redesign are presented in the following section.

6.2   Redesign

The PS10 field layout was redesigned using the free variable method. The optimization was performed by starting with the original PS10 field. Iterations required, on average, 3 hours each. Reported here are the results after 120 iterations. The input values for the PS10 optimization, based on information from Noone *et al.* [20], are summarized in Table 6.1.

Table 6.1: PS10 heliostat field data

Location		Heliostats		Receiver	
Latitude	37°26′N	Count	624	Tower height	115m
Longitude	6°14′W	Height	10m	Aperture width	14m
Site width	800m	Width	12m	Type	Flat
Site length	800m	Geometry	Canted	Tilt	18°

PS10's mirrors are canted; each mirror is shaped such that the focal length of the mirror is equal to its distance from the target. This means that the equation described in the model for determining the image size needs to be modified. The original equation is as follows:

$$D_{\text{image}} = d\lambda + w \qquad (6.1)$$

where d is the distance of the heliostat from the target and w is the largest dimension of the heliostat. Since the mirror is canted, w falls away. Fernández [55] reports that the PS10 field uses Solucar 120 heliostats. The author mentions that these heliostats have an effective error of 2.9 mrad. Incorporating this into the above equation yields:

$$D_{\text{image}} = d(\lambda + 2.9) \qquad (6.2)$$

The original PS10 field was modeled using the above equation to determine image sizes on the receiver. To measure the accuracy of the model, the optical efficiency of the plant at design point was calculated. The mean annual optical efficiency was also calculated. Optical efficiency is the ratio of energy intercepting the receiver to energy incident on the heliostat field [13]. The optical efficiency values were compared to published PS10 data from Noone *et al.* [20]. The values are tabulated in Table 6.2. The higher annual mean value from the Fortran model may result from the fact that the Fortran model does not account for shading caused by the receiver tower.

Table **6.2**: PS10 optical efficiency comparison with Fortran model

	PS10 Data	Fortran Model
Design Point	64.7	63.3
Mean Annual	72.9	79.2

The results of the optimization are summarized in Table 6.3. Because of time constraints, the optimization was only carried out to 120 iterations. At 624 variables, this optimization consisted of 1248 variables and over 700 000 constraints. The Kuhn-Tucker residual [27] at 120 iterations was still fairly large—8.3—indicating that substantial improvements are still possible. The results indicate an improvement of 1.2% on the original PS10 field in annual intercepted energy.

Table **6.3**: PS10 heliostat field data

Field	Annual Intercepted Energy [GWh]	Intercepted Power at Design Point [MW]	Mean Optical Efficiency
Original PS10 Field	111.0	47.6	63.3%
Improved PS10 Field	112.3	48.4	64.4%

Figure 6.2 shows the PS10 field before and after the 120 iterations of the optimization process. Note that these are not converged results. However, with additional

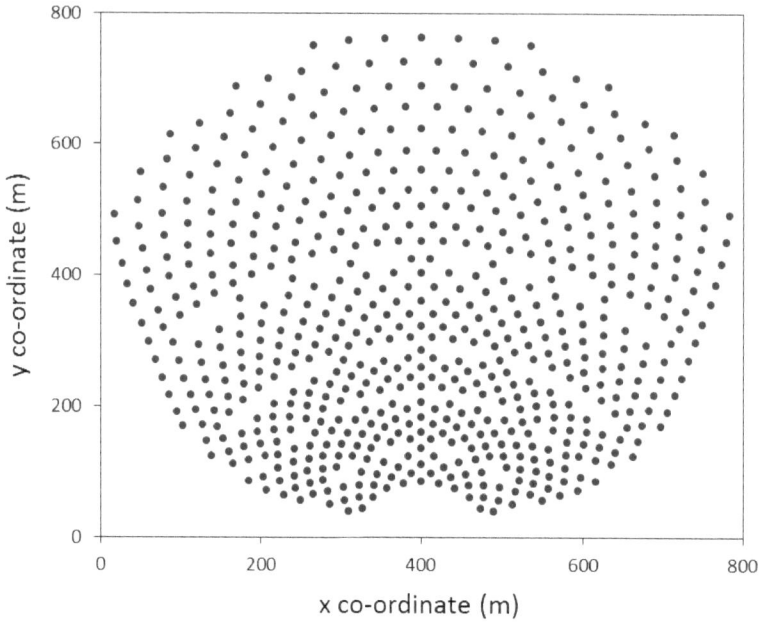

(a) Original PS10 field

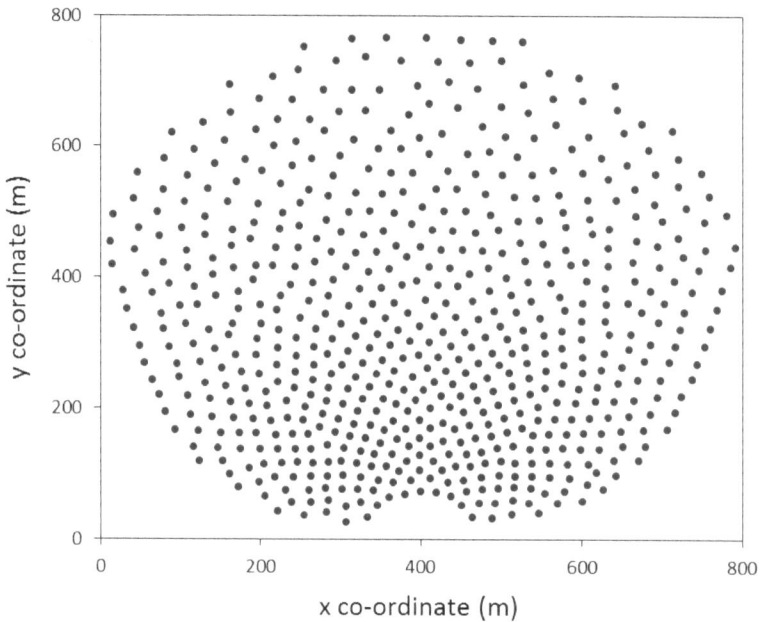

(b) Improved PS10 field

Figure 6.2: PS10 redesign from original PS10

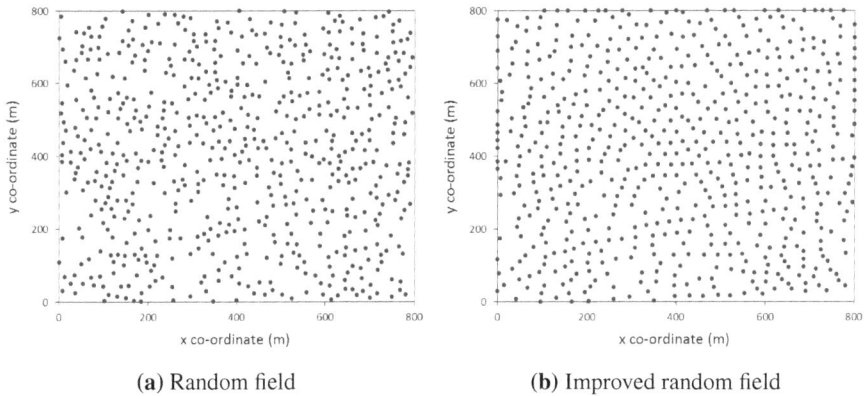

(a) Random field   (b) Improved random field

Figure 6.3: PS10 redesign from random start

iterations further improvements are to be expected. The improvements in the intercepted energy could result in higher revenues for the plant with little or no change in the operation and maintenance costs of the plant.

The fact that the free variable method is able to improve a pattern demonstrates, as Buck [34] has, that optimized patterns do not necessarily result in optimal fields. The free variable method is thus key to obtaining a wholly optimal field.

This optimization was also performed with a random starting point. Figure 6.3 shows the random field with the same specifications as PS10 and the improved random field after 120 iterations of the optimization process. At this stage, the field had not yet reached the performance levels of the original PS10 field. While this result does show a 6% improvement over the random field, it is clear that it is not near an optimal field.

Considering the amount of time taken to reach this point, the case demonstrates that the optimization method is highly time-consuming. The method would require significant refinement before being used for heliostat field layout design and optimization from a random field. However, implementing parallelization can greatly reduce computation time and make this approach viable for field design.

The closeness of the results between the original PS10 field and the field improved from the original field necessitated the need to compare the two fields in a raytracer. The fields were thus simulated in SolTrace. A comparison of the four fields—the original PS10, the field improved from the original, the random field and the field improved from the random field—is shown in Figure 6.4.

Results from SolTrace for the day simulated indicated, as did the Fortran model, a 1.2% improvement by the improved PS10 field from the original PS10 starting point. The random field appears to be superior in intercepted energy during the morning and evening hours. This is most likely due to the fact that the random field is more spread out and thus less susceptible to shading than a more tightly packed field.

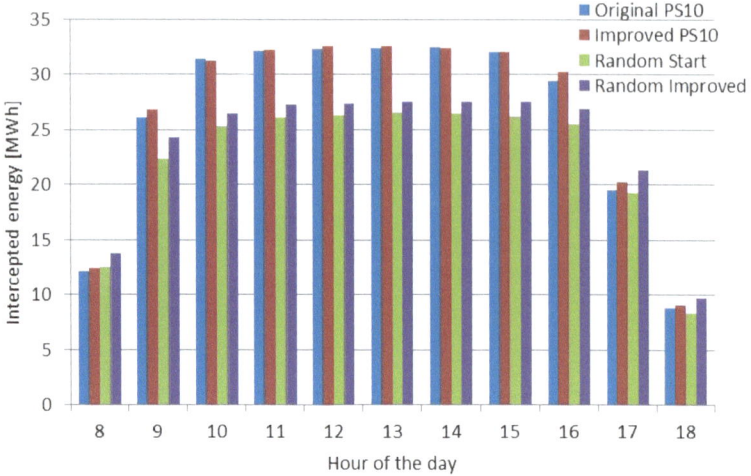

Figure 6.4: A comparison of the PS10 case study fields

With regards to the practicality of the results, the improved PS10 field does not have the same regular rows of heliostats that the original PS10 field does. This means that access roads to heliostats would be affected which, in turn, may have an effect on the maintenance costs of the plant. An improvement of 1.2% in annual intercepted energy would increase the revenues obtained from electricity production, but this needs to be weighed against the negative impact that may be caused by a change in the access roads to the heliostats.

The free variable method allows for additional constraints to be included in the optimization to provide for areas where heliostats cannot be placed such as streams, holes, and restricted areas albeit with an increase in computational time require-ments. Thus, access roads to the heliostats can easily be included in the optimization as constraints. These were neglected here. The tower location, and the boundaries of the site were included, however, demonstrating the functioning of this mechanism.

6.3 Summary

The redesign of the PS10 field demonstrates that the free variable method is capable of producing wholly optimal fields, contrary to the pattern method. The free variable method was able to improve on the original PS10 field in terms of annual intercepted energy. The method is highly time-consuming. This is demonstrated in the redesign performed starting from a random field which proved to be less than satisfactory. Free variable optimization may possibly be considered in commercial CSP for design only if the computational overheads can be addressed adequately.

In the next chapter the findings of this study of heliostat field layout optimization and the free variable method are discussed.

CHAPTER 7

DISCUSSION OF THE APPROACH

From the experience and insight gained from conducting an optimization, this chapter describes some of the key findings in the area of heliostat field layout optimization. A summary of the free variable method explored herein is presented. Thereafter, some of the findings with regards to the characteristics of the analysis method and optimization method used are described.

7.1 The Free Variable Method

The free variable method is a method of heliostat field optimization that follows a more classical approach to optimization. This means iteratively evaluating some function, determining the gradients of the function with respect to each variable and then adjusting each variable to follow the gradient at an optimal step length in the direction of a better function value until a certain objective is achieved [27]. The objective may be the maximization or a minimization of the function value.

In the case of heliostat field optimization, the function may be any of the available field analysis methods such as ray tracing or approximation methods. The objective may be to determine the maximum of the function. As an example, the function may be a calculation of the optical efficiency of the field and the objective may be a maximization of this function. The optimization will keep altering the locations of the heliostats until it can no longer improve the optical efficiency.

To determine the gradient of the function with respect to each variable, a differentiated function is required. If a differentiated function is not available, the gradients may be obtained by finite difference calculations. If, for example, a ray tracer is used as the function, the gradients may be obtained either by finding the partial derivatives of the ray tracer function with respect to each variable or by finite difference calculations where the objective function is evaluated by small perturbations of each variable which, in this case, is each x and y co-ordinate of every heliostat.

At the start of the optimization, the variables may each be assigned a sensible or random value. Any number of equality or inequality constraints can be implemented into the optimization. In heliostat field layout optimization this could include site boundary limitations, distances of heliostats to the tower and distances of heliostats from each other. Furthermore, since heliostats are not limited to a pattern, their motion through the field during optimization is free, which allows for effective consideration of elevation variations within the site as well as discontinuities.

The main drawback of the free variable method of heliostat field optimization is the complexity of the optimization algorithm needed and, as a result, the computational expense. As has been shown in a previous chapter, the free variable method requires the number of optimization variables to be at least double the number of heliostats. It has also been shown that the number of constraints far outnumber the variables. For this reason, sequential approximate optimization was used, which proved to be successful at handling the large number of variables and constraints.

The pattern method and growth method can be done using a much less complex algorithm than the constrained gradient-based algorithm required in the free variable method. This adds to the computational expense of the method. The computational expense is further heightened if a differentiated function is not available; finite difference calculations add significantly to computational expense.

The free variable method is best performed on a high performance computer. The scope for parallelization is vast: obtaining gradient information, analyzing separate heliostats, blocking and shading calculations, and considering the constraints, to name just a few. The free variable method performed on a high performance computer with extensive parallelization appears to be capable of yielding desirable results.

7.2 Field Analysis

Heliostat field layout analysis is required as part of the optimization. This can be done either by ray tracing or by an approximation method. The method described in this book is an approximation method which has the benefit of reduced computational requirements as compared to ray tracing. The approximation model was compared with a ray tracer to determine its accuracy. With the level of accuracy achieved, the model is sufficiently able to drive the optimization towards finding an optimal field.

A disadvantage of the approximation model is that it is constructed in such a way that makes it difficult to differentiate analytically. Thus, during optimization, differentiation by means of calculating finite differences is required. This is a computationally demanding process which adds significantly to the computational time.

7.3 Optimization Attributes

The sequential approximate optimization algorithm used in this approach is a gradient-based method and thus it has the disadvantage of converging to local minima. As has been shown in a previous chapter, the final layout of a heliostat field becomes dependent on the starting point as well as the constraints applied. The final field does usually perform far better than the initial field, but it still cannot be taken as the optimal layout for a given set of system requirements.

It may be possible to determine a global minimum by having multiple starts. A large number of optimization runs could be performed for the same system using different starting conditions. This can be done easily on a high performance cluster. Each optimization can be assigned its own processor. At the end of the exercise the best performing field layout can be chosen.

The time required per iteration in the optimization process increases as the number of heliostats, and hence design variables and constraints, increases. The average time per iteration was determined for three cases each with a different number of heliostats. These are tabulated in Table 7.1. The results indicate that the computational time requirements increase exponentially with the number of heliostats.

Table 7.1: Iteration time requirements for different number of heliostats

Number of heliostats	Average CPU time per iteration [s]
100	120
400	1000
624	5000

7.4 Summary

The free variable method has unique advantages as well as some pitfalls. The free variable method is computationally demanding, yet it appears to be promising with regards to finding optimal heliostat placements. A larger number of heliostats results in exponentially higher computational time. Parallelization and high performance computing may be useful tools for study in this area.

Field analysis which is done by means of an approximation method is not as computationally demanding as common methods in this area, such as ray tracing. It is also less accurate than ray tracing, though it is accurate enough for the optimization approach demonstrated herein.

CHAPTER 8

CONCLUSION

8.1 Findings

This book has demonstrated a classical approach to heliostat field evaluation and lay-out optimization. The free variable method of optimization is a method of optimization that allows heliostats to move freely during optimization to the most optimal point. This book has demonstrated that the method is made possible by the use of a constrained gradient-based optimization algorithm. The results of the method are comparable with results from other methods of heliostat field layout optimization. The free variable method requires further development before being comparable as an alternative method for commercial heliostat field layout optimization. Other findings are summarized below:

- It is possible to construct an accurate heliostat field model for field strength analysis from geometric considerations of the field.

- When starting from a random field the free variable method tends to produce patterns.

- The free variable method is computationally expensive and is best performed on a high performance computer.

- The constraints enforced in the free variable method are large in number and this is a cause for the large computational time requirements.

- Different receiver types result in different field layouts. Optimizations conducted with cylindrical receivers tend to produce spread-out, surround fields while those done with flat receivers produce one-sided fields.

- Each of the three optimization methods for heliostat field layouts provide unique advantages and can be used where fitted. The choice of method depends on the site conditions as well the computational resources available.

8.2   Recommendations

It is recommended that the free variable method remain an open field of research. It appears to be a promising research area. A significant amount of development of this method is needed before it can be applied in industry. Also, it would be useful to develop an analysis model that can be easily differentiated. This could significantly reduce computational time.

8.3   Further work

This section describes suggestions for further work that could be done in this field of research should it be continued. Further work can be done on both the field analysis model and the optimization method.

8.3.1   Analysis model

A useful addition to the analysis model would be the consideration of different shapes of heliostats. This could include circular heliostats, which may reduce the extent of blocking and shading and allow for a more densely packed field.

The analysis model could also be modified to determine the strength of the field in terms of its economic value. A calculation of the levelized energy cost of the plant could thus become the objective of the analysis.

More variables could be added to the optimization by making the analysis model a function of these variables. This may include the tower height and receiver aperture area. In this study, these parameters were taken as constants.

8.3.2   Optimization

The free variable method appears to be dependent on the starting point of the optimization. This means that the optimization has the issue of falling into local minima.

It would be valuable to explore this in more depth and it could be done by having multiple runs of the same problem with different starting points.

A more quantitative analysis of the different methods is needed. It would be useful to test each method with the same problem and to note the differences in performance.

8.3.3 Validation

In the work done herein, validation of the models and results has been done by computer simulation. There is need of physical validation to determine to what extent the models and results compare with real physical systems. This could be done by constructing a model heliostat field and central receiver and determining the intercepted energy for an initial and optimized layout of the heliostats.

APPENDIX A

SAMPLE CALCULATION

A.1 Sample Calculation Problem

For a sample calculation, a simple case is considered consisting of two heliostats. The intercepted energy is calculated for a single hour. The specifications of the case are tabulated in Table A.1. The co-ordinates of the heliostats and receiver tower are given in Table A.2 and depicted in Figure A.1. The goal is to determine the intercepted energy from heliostat 1 at 11am on August 16.

Table A.1: Sample calculation case specifications

Location		Heliostats		Receiver	
Latitude	28°S	Count	2	Tower Height	15m
Longitude	21°E	Height	1m	Type	External cylindrical
Site width	40m	Width	1m	Diameter	1.4m
Site length	40m	Geometry	Flat	Height	1.4m

Table A.2: Sample calculation heliostat and tower co-ordinates

	x	y
Heliostat 1	15	35
Heliostat 2	15	33
Tower	20	0

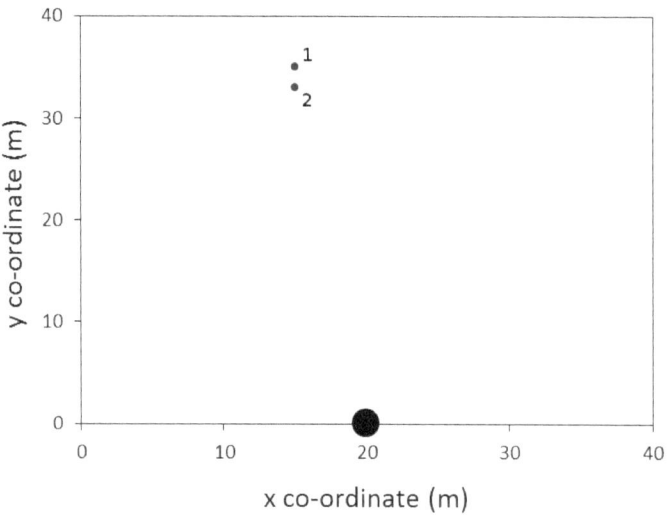

Figure A.1: Sample calculation heliostat field

A.2 Sun Vector

First, the day angle is calculated. August 16 is day 228, thus

$$B = (n-1)\frac{360}{365}$$
$$= (228-1)\frac{360}{365}$$
$$= 223.9°$$

This value is used to determine the equation of time value:

$$E = 229.2(0.000075 + 0.001868\cos B - 0.032077\sin B$$
$$- 0.014615\cos 2B - 0.04089\sin 2B)$$
$$= 229.2(0.000075 + 0.001868\cos(223.9) - 0.032077\sin(223.9)$$
$$- 0.014615\cos(2 \times 223.9) - 0.04089\sin(2 \times 223.9))$$
$$= -4.689$$

Solar time is calculated:

$$\text{Solar time} = \text{Standard time} + [4(L_{loc} - L_{st}) + E]/60$$
$$= 11 + [4(21-30) + (-4.689)]/60$$
$$= 10.32$$

This translates to 10:19:12 since 0.32 of 60 minutes equals 19 minutes and 12 seconds. Using this value, the hour angle can be calculated:

$$\omega = [(\text{Solar time})/24 - 0.5] \times 360$$
$$= [(10.32)/24 - 0.5] \times 360$$
$$= -25.17°$$

Since this value is negative, it accurately shows that the sun is slightly east. The declination angle is then calculated:

$$\delta = 0.006918 - 0.399912\cos(B) + 0.070257\sin(B)$$
$$- 0.006758\cos(2B) + 0.000907\sin(2B)$$
$$- 0.002679\cos(3B) + 0.00148\sin(3B)$$
$$= 0.006918 - 0.399912\cos(223.9) + 0.070257\sin(223.9)$$
$$- 0.006758\cos(2 \times 223.9) + 0.000907\sin(2 \times 223.9)$$
$$- 0.002679\cos(3 \times 223.9) + 0.00148\sin(3 \times 223.9)$$
$$= 0.2442\,\text{rad}$$
$$= 13.99°$$

The plant is located in the Southern hemisphere. The latitude angle at the location is $-28°$. The zenith angle is then

$$
\begin{aligned}
\theta_z &= \cos^{-1}[\cos(\phi) \times \cos(\delta) \times \cos(\omega) + \sin(\phi) \times \sin(\delta)] \\
&= \cos^{-1}[\cos(-28) \times \cos(13.99) \times \cos(-25.17) + \sin(-28) \times \sin(13.99)] \\
&= 48.56°
\end{aligned}
$$

This gives the azimuth angle as

$$
\begin{aligned}
\alpha_z &= 90 - \theta_z \\
&= 90 - 48.56 \\
&= 41.44°
\end{aligned}
$$

The solar azimuth angle is then calculated:

$$
\begin{aligned}
\gamma_s &= \mathrm{sign}(\omega) \left| \cos^{-1} \left(\frac{\cos\theta_z \sin\phi - \sin\delta}{\sin\theta_z \cos\phi} \right) \right| \\
&= \mathrm{sign}(-25.17) \left| \cos^{-1} \left(\frac{\cos(48.56) \sin(-28) - \sin(13.99)}{\sin(48.56) \cos(-28)} \right) \right| \\
&= -146.6°
\end{aligned}
$$

Finally, the three components of the sun vector are determined. The East component is

$$
\begin{aligned}
s_{\mathrm{E}} &= \cos(\alpha_z) \times -\sin(\gamma_s) \\
&= \cos(41.44) \times -\sin(-146.6) \\
&= 0.4127
\end{aligned}
$$

the north component is

$$
\begin{aligned}
s_{\mathrm{N}} &= \cos(\alpha_z) \times -\cos(\gamma_s) \\
&= \cos(41.44) \times -\cos(-146.6) \\
&= 0.6257
\end{aligned}
$$

and the zenith component is

$$
\begin{aligned}
s_z &= \sin(\alpha_z) \\
&= \sin(41.44) \\
&= 0.6619
\end{aligned}
$$

A.3 Target Vector

Since the tower is 15m high, the target's co-ordinates are (20, 0, 15). The target vector for heliostat 1 then is

$$\mathbf{T}_1 = \begin{bmatrix} x_T - x_1 \\ y_T - y_1 \\ z_T - z_1 \end{bmatrix} = \begin{bmatrix} 20 - 15 \\ 0 - 35 \\ 15 - 0 \end{bmatrix} = \begin{bmatrix} 5 \\ -35 \\ 15 \end{bmatrix}$$

The magnitude of this vector is

$$||\mathbf{T}_1|| = \sqrt{(5^2 + (-35)^2 + 15^2)}$$
$$= 38.41$$

thus, the unit vector is

$$\mathbf{t}_1 = \begin{bmatrix} 5/38.41 \\ -35/38.41 \\ 15/38.41 \end{bmatrix} = \begin{bmatrix} 0.1302 \\ -0.9113 \\ 0.3906 \end{bmatrix}$$

Similarly, for heliostat 2, the target vector is

$$\mathbf{T}_2 = \begin{bmatrix} 5 \\ -33 \\ 15 \end{bmatrix}$$

and its unit vector is

$$\mathbf{t}_2 = \begin{bmatrix} 0.1366 \\ -0.9018 \\ 0.4099 \end{bmatrix}$$

At this point it is necessary to change from the x-y-z co-ordinate system to the i-j-k co-ordinate system. This is because the sun vector uses this convection. Any further calculations need to be in the same co-ordinate system.

In the x-y-z, x and y are positive west and south respectively. In the i-j-k system, i and j are positive east and north. Thus, to change from the to the i-j-k system, the signs of the x and y values of the vectors need to be changed. The target unit vectors become

$$\mathbf{t}_1 = \begin{bmatrix} -0.1302 \\ 0.9113 \\ 0.3906 \end{bmatrix} \text{ and } \mathbf{t}_2 = \begin{bmatrix} -0.1366 \\ 0.9018 \\ 0.4099 \end{bmatrix}$$

A.4   Heliostat Normal

The heliostat normal for heliostat 1 is

$$\mathbf{N}_1 = \mathbf{s} + \mathbf{t}_1$$

$$= \begin{bmatrix} 0.4127 + (-0.1302) \\ 0.6257 + 0.9113 \\ 0.6619 + 0.3906 \end{bmatrix} = \begin{bmatrix} 0.2825 \\ 1.537 \\ 1.052 \end{bmatrix}$$

The unit vector for this normal vector is

$$\mathbf{n}_1 = \frac{\mathbf{N}_1}{||\mathbf{N}_1||}$$

$$= \begin{bmatrix} 0.1499 \\ 0.8158 \\ 0.5586 \end{bmatrix}$$

Similarly, the unitized heliostat normal of heliostat 2 is

$$\mathbf{n}_2 = \begin{bmatrix} 0.1464 \\ 0.8098 \\ 0.5682 \end{bmatrix}$$

It will be seen later that these two heliostats are close enough to each other to cause blocking. In blocking, the assumption is made that heliostats close enough to each other to cause blocking have the same orientation. The similarity of the normal vectors of the two heliostats illustrates the validity of this assumption.

A.5   Cosine Efficiency

The cosine efficiency of heliostat 1 is

$$\eta_c = \mathbf{s} \cdot \mathbf{n}_1$$
$$= 0.4127 \times 0.1499 + 0.6257 \times 0.8158 + 0.6619 \times 0.5586$$
$$= 0.9421$$

A.6   Attenuation Efficiency

To calculate the attenuation efficiency of heliostat 1, first the distance of the heliostat from the target is calculated. This is done using the x-y-z co-ordinates of the heliostat and the tower:

$$d_T = ||(x_1, y_1, z_1) - (x_T, y_T, z_T)||$$
$$= ||(15, 35, 0) - (20, 0, 15)||$$
$$= 38.41$$

The attenuation efficiency is then

$$\eta_{a_i} = 0.99321 - 0.0001176 \cdot d_T + 1.97 \times 10^{-8} \cdot d_T^2$$
$$= 0.99321 - 0.0001176 \cdot 38.41 + 1.97 \times 10^{-8} \cdot 38.41^2$$
$$= 0.9887$$

A.7 Spillage Efficiency

Using the distance, d_T, determined in the previous calculation, the size of the re-flected image at the receiver is determined:

$$D_{\text{image}} = d_T \lambda + w$$
$$= (38.41)(9.3 \times 10^{-3}) + 1$$
$$= 1.357 \, \text{m}$$

To determine the angle, α, at which the image intercepts the receiver, the horizontal distance of the heliostat to the tower is first determined:

$$d_{xy} = ||(x_T, y_T, 0) - (x_1, y_1, 0)||$$
$$= ||(20, 0, 0) - (15, 35, 0)||$$
$$= 35.36 \, \text{m}$$

Using this distance, α is calculated to be:

$$\alpha = \sin^{-1}\left(\frac{d_{xy}}{d_T}\right)$$
$$= \sin^{-1}\left(\frac{35.36}{38.41}\right)$$
$$= 67.01°$$

This angle is used to calculate the vertical lengthening of the elliptical image:

$$L_v = \frac{D_{\text{image}}}{\sin \alpha}$$
$$= \frac{1.357}{\sin(67.01)}$$
$$= 1.474 \, \text{m}$$

To determine the extent of spillage, the total area of the actual image is calculated:

$$A_{\text{total}} = \frac{\pi}{4} L_v D_{\text{image}}$$
$$= \frac{\pi}{4}(1.474)(1.357)$$
$$= 1.571\,\text{m}^2$$

The total ineffective area can then be calculated. Since the length of the image in the horizontal direction does not exceed the diameter of the receiver, only the ineffective area caused by the vertical length needs to be determined:

$$A_{\text{ineffective}} = \frac{(L_v - H_{\text{Receiver}}) \cdot D_{\text{image}}}{1.284}$$
$$= \frac{(1.474 - 1.4) \cdot 1.357}{1.284}$$
$$= 0.07821\,\text{m}^2$$

The effective area is then:

$$A_{\text{effective}} = A_{\text{total}} - A_{\text{ineffective}}$$
$$= 1.571 - 0.07821$$
$$= 1.493\,\text{m}^2$$

Thus, the spillage efficiency is

$$\eta_{sp} = \frac{A_{\text{effective}}}{A_{\text{total}}}$$
$$= \frac{1.493}{1.571}$$
$$= 0.9502$$

A.8   Blocking Efficiency

To determine whether or not blocking will take place, the necessary vectors are to be determined. The target vector is the same as what was calculated above:

$$\mathbf{T} = \begin{bmatrix} 5 \\ -35 \\ 15 \end{bmatrix}$$

The vector $\mathbf{R}$, which points from the heliostat under consideration, heliostat 1, to the potentially blocking heliostat, heliostat 2, is

$$\mathbf{R} = \begin{bmatrix} x_2 - x_1 \\ y_2 - y_1 \\ z_2 - z_1 \end{bmatrix}$$

$$= \begin{bmatrix} 15 - 15 \\ 33 - 35 \\ 0 - 0 \end{bmatrix}$$

$$= \begin{bmatrix} 0 \\ -2 \\ 0 \end{bmatrix}$$

The shortest distance, d, from heliostat 2 to the target vector line of heliostat 1 is then

$$d = \frac{|\mathbf{T} \times \mathbf{R}|}{|\mathbf{T}|}$$

$$= \frac{\left| \begin{bmatrix} 5 \\ -35 \\ 15 \end{bmatrix} \times \begin{bmatrix} 0 \\ -2 \\ 0 \end{bmatrix} \right|}{\left| \begin{bmatrix} 5 \\ -35 \\ 15 \end{bmatrix} \right|}$$

$$= 0.8234 \, \text{m}$$

Next, the critical distance, d_c, is determined:

$$d_c = (H_w^2 + H_h^2)^{\frac{1}{2}}$$
$$= (1^2 + 1^2)^{\frac{1}{2}}$$
$$= 1.414 \, \text{m}$$

Since the distance, d, is smaller than the critical distance, d_c, there is a potential for blocking. It is now necessary to determine whether or not heliostat 2 is closer to the target than heliostat 1. To do this, the scalar parameter, t, of the equation of the target vector line is determined:

$$t = \frac{\mathbf{R} \cdot \mathbf{T}}{|\mathbf{T}|^2}$$

$$= \frac{\begin{bmatrix} 0 \\ 2 \\ 0 \end{bmatrix} \cdot \begin{bmatrix} 5 \\ -35 \\ 15 \end{bmatrix}}{\left| \begin{bmatrix} 5 \\ -35 \\ 15 \end{bmatrix} \right|^2}$$

$$= 0.04746$$

Since this parameter is positive, heliostat 2 is indeed closer to the target than heliostat 1 and will cause blocking. The extent of blocking can now be calculated. For this calculation, the heliostat center becomes the origin of the local co-ordinate system.

To determine the extent of blocking, the two heliostat faces must be discretized, and lines need to be extended from the discretization points. Firstly, the point separation distances are determined along the height

$$\delta_h = \frac{H_h}{3}$$
$$= \frac{1}{3}$$
$$= 0.3333 \, \text{m}$$

and the width

$$\delta_w = \frac{H_w}{3}$$
$$= \frac{1}{3}$$
$$= 0.3333 \, \text{m}$$

With the center of the heliostat at $(0,0,0)$, each node will have its own unique co-ordinates determined by the orientation of the heliostat. For this example, the co-ordinates of a node to the right of the central node will be determined. The heliostat is treated as a plane in 3-dimensional space. The equation of a plane is simply

$$ax + by + cz = 0$$

where $\langle a, b, c \rangle$ is a vector perpendicular to the plane. This is the normal vector that was calculated in a previous step. The vector, which can be denoted $\mathbf{P}$ in the plane of the heliostat and parallel to the horizontal, can be determined by setting $z = 0$ in the equation of the plane and giving y a positive value, of, for example, 1. Using the heliostat as the local x-y-z co-ordinate system, x can then be calculated as

$$ax = -by$$
$$x = -by/a$$
$$= -(-0.8158)(1)/(-0.1499)$$
$$= -5.442$$

Thus, the vector in the plane of the heliostat and parallel to the horizontal is $P = \langle -5.442, 1, 0 \rangle$. This vector can be used to determine the co-ordinates of the nodes left and right of the center. This vector can be unitized by dividing each component by the magnitude of the vector, giving $p = \langle -0.9837, 0.1808, 0 \rangle$.

The next vector needed is a vector, q, perpendicular to p and pointing towards the top of the heliostat. This vector can be used to determine the co-ordinates of nodes above and below the central node; thus all the nodes can be mapped out. Since q is perpendicular to p and the normal vector, n_1, q is simply the cross product of n_1 and q:

$$q = n_1 \times p$$

$$= \begin{bmatrix} -0.1499 \\ -0.8158 \\ 0.5586 \end{bmatrix} \times \begin{bmatrix} -0.9837 \\ 0.1808 \\ 0 \end{bmatrix}$$

$$= \begin{bmatrix} -0.1010 \\ -0.5489 \\ -0.8288 \end{bmatrix}$$

The co-ordinates of all 9 nodal points can be determined using the vectors p and q. To determine the co-ordinates of node in the top right corner of the heliostat face, using the center of the heliostat as the origin, a unit of separation is added to the corresponding component along the appropriate vector. The top right node is a single step to the right plus a single step upwards of the center. Its co-ordinates in the local co-ordinate system are:

$$\begin{bmatrix} x \\ y \\ z \end{bmatrix} = \begin{bmatrix} 0 \\ 0 \\ 0 \end{bmatrix} + 1 \times \delta_w p + 1 \times \delta_h q$$

$$= \begin{bmatrix} 0 \\ 0 \\ 0 \end{bmatrix} + 1 \times (0.3333) \begin{bmatrix} -0.9837 \\ 0.1808 \\ 0 \end{bmatrix} + 1 \times (0.3333) \begin{bmatrix} -0.1010 \\ -0.5489 \\ -0.8288 \end{bmatrix}$$

$$= \begin{bmatrix} -0.3641 \\ -0.1227 \\ -0.2762 \end{bmatrix}$$

Similarly, the local co-ordinates of all the nodes on the heliostat surface can be determined.

Since the assumption is made that the two heliostats have the same orientation, the vectors $\mathbf{p}$ and $\mathbf{q}$ can be used to determine nodal points on heliostat 2 as well. To do this, the origin of heliostat 2 relative to heliostat 1 is used as the starting point as opposed to $\langle 0, 0, 0 \rangle$ used in the calculation above. To determine the co-ordinates of the center of heliostat 2 relative to heliostat 1, the global co-ordinates of heliostat 2 are subtracted from the global co-ordinates of heliostat 1. Essentially, the co-ordinates have the same values as the components of the vector $\mathbf{R}$ calculated for the blocking potential check.

After finding the co-ordinates of all the points on both heliostats, the next step in determining the extent of blocking is to extend lines from each of the nodes on heliostat 1 and to determine whether or not they intersect heliostat 2. For this example, a line from the central node of heliostat 1 and the central node of heliostat 2 will be used.

The calculation is basically a line-plane intersection calculation where the plane is the blocking heliostat, heliostat 2. The value of scalar parameter, t, of the line from heliostat 1 at the point of intersection with heliostat 2 needs to be determined. Weisstein [45] shows that, given three points on a plane, $\mathbf{x_1}, \mathbf{x_2}, \mathbf{x_3}$, and two points on a line, $\mathbf{x_4}$ and $\mathbf{x_5}$, the scalar parameter, t, along the line at the point of intersection of the line and the plane can be determined as follows:

$$t = -\frac{\begin{vmatrix} 1 & 1 & 1 & 1 \\ x_1 & x_2 & x_3 & x_4 \\ y_1 & y_2 & y_3 & y_4 \\ z_1 & z_2 & z_3 & z_4 \end{vmatrix}}{\begin{vmatrix} 1 & 1 & 1 & 0 \\ x_1 & x_2 & x_3 & x_5 - x_4 \\ y_1 & y_2 & y_3 & y_5 - y_4 \\ z_1 & z_2 & z_3 & z_5 - z_4 \end{vmatrix}}$$

Since all the nodes are in the plane of the heliostats, any 3 of these can be used, along with any two points on the line. The first point on the line can be the node on the heliostat, and the second point can be determined using the target vector:

$$\mathbf{x_5} = \mathbf{x_4} + \mathbf{T}$$

Substituting the co-ordinates of all the points needed to solve for t, t is found to be 0.04518. This value for t can now be used to determine the co-ordinates of the intersection point using the equation of the line:

$$\mathbf{x}' = \mathbf{x}_4 + \mathbf{T}t$$

$$= \begin{bmatrix} 0 \\ 0 \\ 0 \end{bmatrix} + \begin{bmatrix} 5 \\ -35 \\ 15 \end{bmatrix} 0.04518$$

$$= \begin{bmatrix} 0.2259 \\ -1.581 \\ 0.6777 \end{bmatrix}$$

This point can now be used to determine the proximity of the intersection point from the central node of heliostat 2:

$$d = ||\mathbf{x}' - \mathbf{R}||$$

$$= \left\| \begin{bmatrix} 0.2259 \\ -1.581 \\ 0.6777 \end{bmatrix} - \begin{bmatrix} 0 \\ -2 \\ 0 \end{bmatrix} \right\|$$

$$= 0.8280$$

A critical distance is now needed to determine whether or not this line is in close proximity to the node. This critical distance can be taken to be half the hypotenuse of the two separation distances of the nodes:

$$d_c = \left(\sqrt{\delta_h^2 + \delta_w^2} \right) / 2$$

$$= \left(\sqrt{0.3333^2 + 0.3333^2} \right) / 2$$

$$= 0.2357 \, \text{m}$$

Since the distance of the intersection point is further than the critical distance from the node, it cannot be confirmed that this line intersects heliostat 2. However, it might be close enough to one of the other nodes on heliostat 2. This calculation must therefore be done with all of the nodes on heliostat 2. If the intersection point comes within the critical distance, d_c, of any of the nodes, the line from heliostat 1 intersects heliostat 2.

After doing the calculation for all combinations of nodes, it is found that the number of lines from heliostat 1 intersecting heliostat 2 is 6. The extent of blocking can now be determined:

$$\eta_{b_1} = \frac{\text{Number of intersecting lines}}{9}$$

$$= \frac{6}{9}$$

$$= 0.6667$$

A.9 Shading Efficiency

The shading efficiency calculation proceeds in the same manner that the blocking calculation does. The only difference is that the sun vector, **S**, is used for the equations of all the lines instead of the target vector.

After doing the calculation for shading, it is found that heliostat 2 does not shade heliostat 1. Thus the shading efficiency at this hour is:

$$\eta_{s_1} = 1$$

A.10 Intercepted Energy

Having obtained all the efficiencies for heliostat 1, the intercepted energy from heliostat 1 can be calculated. The DNI at this hour is $955 \, \text{W/m}^2$. Thus, the intercepted energy from heliostat 1 for this hour is

$$
\begin{aligned}
I &= A \times \text{DNI} \times \eta_{c_1} \eta_{a_1} \eta_{sp_1} \eta_{b_1} \eta_{s_1} \\
&= (1 \times 1) \times (955) \times (0.9421)(0.9887)(0.9502)(0.6667)(1) \\
&= 563.5 \, \text{Wh}
\end{aligned}
$$

APPENDIX B

COMPUTER CODE

B.1 Fortran Code

```
 1    program test1
 2
 3    integer u, fdata(12), i
 4    parameter (n = 4, u = 20)
 5    double precision Field1(n), f, ruser(3,600), DNIN, OptE,
 6   +                 d4,d5
 7
 8    !Integers Only!
 9    fdata(1) = 1      !Mirror type (1 = flat; 2 = curved)
10    fdata(2) = 1      !Field direction (1 = south field; 2 =
         north field)
11    fdata(3) = 15     !Tower Height
12    fdata(4) = 1      !Heliostat Height (manually changed
         in Objective function if non-integer)
13    fdata(5) = 1      !Heliostat Width (manually changed in
         Objective function if non-integer)
```

```
14        fdata(6)  = n/2        !Number of heliostats
15        fdata(7)  = 192        !Number of hours
16        fdata(8)  = 2          !Receiver Width
17        fdata(9)  = 20                  !Receiver x-co-ordinate
18        fdata(10) = 0                   !Receiver y-co-ordinate
19        fdata(11) = 1          !Receiver type (1 = external, 2 =
            cavity)
20        fdata(12) = 1          !Pylon Height
21
22  !     Starting Point
23
24
25  C       open(u, FILE = 'Field_2.txt', STATUS = 'OLD')
26  C           do 10 i = 1, n
27  C               read(u,*) Field1(i)
28  c10         continue
29  C       close(u)
30
31  !                       x-co-ordinates
32                          Field1(1) = 15
33                          Field1(2) = 15
34  C                       Field1(3) = 20
35  C
36  !                       y-co-ordinates
37                          Field1(n/2 + 1) = 33
38                          Field1(n/2 + 2) = 35
39  C                       Field1(n/2 + 3) = 130
40
41        d4 = fdata(4)
42        d5 = fdata(5)
43
44        call dsinput(ruser,fdata)
45
46        call heliostat(Field1,n,ruser,fdata,f)
47
48        write (*,*) f
49
50        DNIN = 0
51
52  C     Optical Efficiency ----------------
53        do 134 i = 1,192
54           DNIN = ruser(1,i) + DNIN
55  134   continue
56        OptE = -f/(DNIN*d4*d5*n/2)
57        write (*,*) 'Optical Efficiency: ', OptE
58
59
60        stop
61        end
```

```fortran
62
63
64          subroutine heliostat(Field,n,ruser,fdata,Energy)
65   c Variables
66          integer    fdata(*), n, n2, i, j, k, l, q,
67   c Blocking and Shading
68          +              t1,q1,r1,pmatx(25),pmaty(25),flag1,
69   c Other
70          +              flag, Tower(3), flag2,
71   c Field data
72          +              d1, d2, d3,          d6, d7, d8, d9, d10, d11,
            d12
73   c
74          double precision Area(fdata(7)), d4, d5,
75   c Angles
76          +              Theta1, Theta2,
77   c Efficiencies
78          +              n_at(fdata(6)), n_s, n_b, n_c, n_sp(fdata(6)),
            n_tot,
79   c Vectors
80          +              U1(3), Target1(3), Normal(3), t_mag, u_mag,
81          +              n_mag, s_mag, Txy(3), Nxy(3), Sxy(3),
82   c Other
83          +              db(4), ds(3), xa(3), sundist,
84          +              b_h, b_w, s_h, s_w, dist2, diag, temp1, temp2,
            temp3,
85   c Spillage
86          +              distance(fdata(6)), distance2(fdata(6)), image
            , image2,
87          +              alpha1,
88   c Blocking and Shading
89          +              rc(3), t,ff,this(25,3),that(25,3),thisp(3),
            dist,
90          +                                   thatp(3), thisu(3),
            thatu(3), tt_mag, dcw, dch, dcd,
91          +              Td(3), r2,det1, det2, pt,
92   c Other
93          +              pi, Energy, Field(*), ruser(3,600), Fieldz(n
            /2),
94          +              MAT(4,4),MAT2(4,4), x5(3)
95
96   c Field Data
97          d1 = fdata(1)    !Mirror type (1 = flat; 2 = curved)
98          d2 = fdata(2)    !Field direction (1 = south field; 2 =
            north field)
99          d3 = fdata(3)    !Tower Height
100         d4 = fdata(4)    !Heliostat Height
101         d5 = fdata(5)    !Heliostat Width
102         d6 = fdata(6)    !Number of heliostats
```

```
103       d7 = fdata(7)     !Number of hours
104       d8 = fdata(8)     !Receiver Width/Height
105       d9 = fdata(9)     !Receiver x-co-ordinate
106       d10= fdata(10)    !Receiver y-co-ordinate
107       d11= fdata(11)    !Receiver type (1 = external, 2 = cavity
             )
108       d12= fdata(12)    !Pylon height
109
110   c Non-integer dimesioned heliostats
111   c       d4 = 1.4
112   c       d5 = 1.4
113       ff = 0
114   c ----Number of Heliostats
115       n2 = n/2
116       pi= 3.14159265359d0
117
118   c ----Heliostat property
119       diag = d4**2 + d5**2
120       diag = sqrt(diag)
121       dcw = d5/5
122       dch = d4/5
123       dcd = sqrt(dcw**2 + dch**2)
124
125
126   c ----Heliostat Matrix for blocking and shading points
127       r1 = 0
128       do t1 = 1,5
129          do q1 = 1,5
130             r1 = r1 + 1
131             pmatx(r1)=-4+q1
132          end do
133       end do
134
135       r1 = 0
136       do t1 = 1,5
137          do q1 = 1,5
138             r1 = r1 + 1
139             pmaty(r1)=4-t1
140          end do
141       end do
142
143   c Objective Function------------
144
145       Tower(1) = d9
146       Tower(2) = d10
147       Tower(3) = d3
148
149       Energy = 0
150
```

```fortran
151          call topology(Field,Fieldz,n2)
152
153          flag2 = 0 !indication as to whether attenuation and
                 spillage have been calculated
154
155
156          do 1000 i = 1, d7
157            Area(i) = 0
158          if (ruser(1,i) .gt. 0) then !DNI > 0
159
160 c Sun vector for this hour in ijk format
161          U1(1) = ruser(2,i)
162          U1(2) = ruser(2,i+d7)
163          U1(3) = ruser(2,i+d7*2)
164
165          u_mag = U1(1)**2 + U1(2)**2 + U1(3)**2
166
167          U1(1) = U1(1)/u_mag
168          U1(2) = U1(2)/u_mag
169          U1(3) = U1(3)/u_mag
170
171
172 !Sun Vector in xy co-ordinates
173          Sxy(1) = -U1(1)
174          Sxy(2) = -U1(2)
175          Sxy(3) =  U1(3)
176
177 c For each heliostat
178          do 100 j = 1,n2
179 c Target Vector in ijk co-ordintates
180            Target1(1) = -(Tower(1) - Field(j))
181            Target1(2) = -(Tower(2) - Field(j+n2))
182            Target1(3) = Tower(3) - Fieldz(j)
183
184            t_mag = sqrt(Target1(1)**2 + Target1(2)**2 +
185      +            Target1(3)**2)
186
187 c Distance from heliostat to tower
188 c            distance = sqrt(Target1(1)**2 + Target1(2)**2 +
189 c      +            Target1(3)**2)
190
191 c Normal unit vector
192            Normal(1) = (Target1(1)/t_mag + U1(1))
193            Normal(2) = (Target1(2)/t_mag + U1(2))
194            Normal(3) = (Target1(3)/t_mag + U1(3))
195            n_mag = sqrt(Normal(1)**2 + Normal(2)**2 +
196      +            Normal(3)**2)
197
198            Normal(1) = Normal(1)/n_mag
```

```fortran
199              Normal(2) = Normal(2)/n_mag
200              Normal(3) = Normal(3)/n_mag
201
202    c Cosine ------------------------
203              n_c = U1(1)*Normal(1) + U1(2)*Normal(2) + U1(3)*
204         +          Normal(3)
205
206
207              if (flag2 .eq. 0) then !attenuation and spillage
                     need only be calculated once
208    c Distance from heliostat to tower
209              distance(j) = sqrt((Tower(1) - Field(j))**2 + (
                     Tower(2) -
210         +              Field(j+n2))**2 + (Tower(3) - Fieldz(j))
                  **2)
211
212    c Attenuation ------------------
213              n_at(j) =  0.99321 - 0.0001176*distance(j) +
214         +              1.97*(10**(-8))*distance(j)**2
215
216    c Spillage --------------------
217
218              if (d1 .eq. 1) then
219                  image = distance(j)*9.3e-3 + d5
220              endif
221
222              if (d1 .eq. 2) then
223                  image = distance(j)*9.3e-3
224              endif
225
226              !External Receiver------------
227              if (d11 .eq. 1) then
228                  distance2(j) = sqrt((Tower(1) - Field(j))**2 +
229         +                  (Tower(2)-Field(j+n2))**2)
230                  alpha1 = asin(distance2(j)/distance(j))
231                  image2 = image/sin(alpha1)
232
233                  n_sp(j) = 1
234
235                  if (image2 .gt. d8) then
236                      temp1 = image2*image*pi/4
237                      temp2 = temp1 - (image2-d8)*image/1.284
238                      temp2 = temp2/temp1
239                      n_sp(j) = n_sp(j)*temp2
240                  endif
241
242                  if (image .gt. d8) then
243                      temp1 = image2*image*pi/4
244                      temp2 = temp1 - (image - d8)*image2/1.284
```

```
245                 temp2 = temp2/temp1
246                 n_sp(j) = n_sp(j)*temp2
247               endif
248
249             if (n_sp(j) .gt. 1) then
250               n_sp(j) = 1
251             endif
252           endif !External receiver--------
253
254           !Cavity receiver----------------
255           if (d11 .eq. 2) then
256             n_sp(j) = 1
257
258             distance2(j) = ((-1)**d2)*(Tower(2)-Field(j+n2
                  )) !negative for south field, positive for
                  north
259             alpha1 = asin(distance2(j)/distance(j))
260
261             if (alpha1 .le. 0) then
262               n_sp(j) = 0
263               image2 = 0
264               image = 0
265             elseif (alpha1 .gt. 0) then
266               image2 = image/sin(alpha1)
267             endif
268
269             if (image2 .gt. d8) then
270               temp1 = image2*image*pi/4
271               temp2 = temp1 - (image2-d8)*image/1.284
272               temp2 = temp2/temp1
273               n_sp(j) = n_sp(j)*temp2
274             endif
275
276             if (image .gt. d8) then
277               temp1 = image2*image*pi/4
278               temp2 = temp1 - (image - d8)*image2/1.284
279               temp2 = temp2/temp1
280               n_sp(j) = n_sp(j)*temp2
281             endif
282
283             if (n_sp(j) .gt. 1) then
284               n_sp(j) = 1
285             endif
286           endif !Cavity receiver----------
287
288           !PS10 receiver----------------
289           if (d11 .eq. 3) then
290             n_sp(j) = 1
291
```

```
292             distance2(j) = ((-1)**d2)*(Tower(2)-Field(j+n2
                   )) !negative for south field, positive for
                   north
293             beta1 = atan(Tower(3)/(abs(Tower(2) - Field(j)
                   )))
294
295             alpha1 = asin(distance2(j)/distance(j))
296             alpha1 = alpha1 + (18*pi/180)*sin(beta1) !PS10
                   's receiver slants 18deg downwards
297
298             if (alpha1 .le. 0) then
299               n_sp(j) = 0
300               image2 = 0
301               image = 0
302             elseif (alpha1 .gt. 0) then
303               image2 = image/sin(alpha1)
304             endif
305
306             if (image2 .gt. d8) then
307               temp1 = image2*image*pi/4
308               temp2 = temp1 - (image2-d8)*image/1.284
309               temp2 = temp2/temp1
310               n_sp(j) = n_sp(j)*temp2
311             endif
312
313             if (image .gt. d8) then
314               temp1 = image2*image*pi/4
315               temp2 = temp1 - (image - d8)*image2/1.284
316               temp2 = temp2/temp1
317               n_sp(j) = n_sp(j)*temp2
318             endif
319
320             if (n_sp(j) .gt. 1) then
321               n_sp(j) = 1
322             endif
323           endif !PS10 receiver----------
324
325
326         endif
327
328 c Blocking and Shading-------------
329
330         !Target Vector in xy co-ordinates
331         Txy(1) = -Target1(1)
332         Txy(2) = -Target1(2)
333         Txy(3) =  Target1(3)
334
335         !Generate Points on this heliostat
336         !Normal Vector in xy co-ordinates
```

```fortran
337          Normal(1) = -Normal(1)
338          Normal(2) = -Normal(2)
339          !Normal(3) = Normal(3)
340
341          !Plane Vector
342          thisp(1) = -Normal(2)*1/(Normal(1))
343          thisp(2) = 1
344          thisp(3) = 0
345          tt_mag = sqrt(thisp(1)**2 + thisp(2)**2 + thisp(3)
                   **2)
346          thisp(1) = thisp(1)/tt_mag
347          thisp(2) = thisp(2)/tt_mag
348          thisp(3) = thisp(3)/tt_mag
349
350          !Up Vector
351          thisu(1) =   Normal(2)*thisp(3) - Normal(3)*thisp(2)
352          thisu(2) = -(Normal(1)*thisp(3) - Normal(3)*thisp(1)
                   )
353          thisu(3) =   Normal(1)*thisp(2) - Normal(2)*thisp(1)
354
355          !Points on this heliostat
356          do t1 = 1,25
357            do q1 = 1,3
358               this(t1,q1) = 0 + dcw*pmatx(t1)*thisp(q1) +
359     +                    dch*pmaty(t1)*thisu(q1)
360
361
362            end do
363          end do
364
365          n_b = 1
366          n_s = 1
367
368          do 40 k = 1,n2
369          if (k .ne. j) then
370
371           !Relative x and y co-ordinates
372           rc(1) = Field(k) - Field(j)
373           rc(2) = Field(k+n2) - Field(j+n2)
374           rc(3) = Fieldz(k) - Fieldz(j)
375
376  c Blocking ----------------------
377
378           !Point to a line (distance between target vector
                   line and heliostat k)
379           xa(1) = -rc(1)
380           xa(2) = -rc(2)
381           xa(3) = -rc(3)
382
```

```
383          db(1) =     Txy(2)*xa(3) - Txy(3)*xa(2)
384          db(2) =   -(Txy(1)*xa(3) - Txy(3)*xa(1))
385          db(3) =     Txy(1)*xa(2) - Txy(2)*xa(1)

387          temp1 = db(1)**2 + db(2)**2 + db(3)**2
388          temp1 = sqrt(temp1)
389          temp2 = Txy(1)**2 + Txy(2)**2 + Txy(3)**2
390          temp2 = sqrt(temp2)

392          db(4) = temp1/temp2

394          !is this heliostat close enough to the other
                 heliostat to potentially block?
395          if (db(4) .lt. diag) then
396          !is "that" heliostat in front of "this"?
397             temp1 = Tower(1) - Field(k)
398             temp2 = Tower(2) - Field(k+n2)
399             temp3 = Tower(3) - Fieldz(k)
400             dist2 = sqrt(temp1**2 + temp2**2 + temp3**2)


403             if (dist2 .lt. distance(j)) then
404                !Generate points on "that" heliostat
405                do t1 = 1,25
406                   do q1 = 1,3
407                      that(t1,q1) = rc(q1) + dcw*pmatx(t1)*
                            thisp(q1)
408       +                           + dch*pmaty(t1)*thisu(q1)
409                   end do
410                end do


413             MAT(1,1) = 1d0
414             MAT(1,2) = 1d0
415             MAT(1,3) = 1d0
416             MAT(1,4) = 1d0

418             MAT(2,1) = that(1,1) !x1
419             MAT(2,2) = that(5,1) !x2
420             MAT(2,3) = that(25,1) !x3
421  !          MAT(2,4) = 0 !x4 Specific to the point on "
     this" heliostat

423             MAT(3,1) = that(1,2) !y1
424             MAT(3,2) = that(5,2) !y2
425             MAT(3,3) = that(25,2) !y3
426  !          MAT(3,4) = 0 !y4 Specific to the point on "
     this" heliostat
```

```
428            MAT(4,1) = that(1,3) !z1
429            MAT(4,2) = that(5,3) !z2
430            MAT(4,3) = that(25,3) !z3
431  !         MAT(4,4) = 0 !z4 Specific to the point on "
       this" heliostat
432
433
434            MAT2(1,1) = 1d0
435            MAT2(1,2) = 1d0
436            MAT2(1,3) = 1d0
437            MAT2(1,4) = 0d0
438
439            MAT2(2,1) = that(1,1) !x1
440            MAT2(2,2) = that(5,1) !x2
441            MAT2(2,3) = that(25,1) !x3
442  !         MAT2(2,4) = 0 !x5-x4 Specific to the point
       on "this" heliostat
443
444            MAT2(3,1) = that(1,2) !y1
445            MAT2(3,2) = that(5,2) !y2
446            MAT2(3,3) = that(25,2) !y3
447  !         MAT2(3,4) = 0 !y5-y4 Specific to the point
       on "this" heliostat
448
449            MAT2(4,1) = that(1,3) !z1
450            MAT2(4,2) = that(5,3) !z2
451            MAT2(4,3) = that(25,3) !z3
452  !         MAT2(4,4) = 0 !z5-z4 Specific to the point
       on "this" heliostat
453
454
455            flag1 = 0
456            r2 = 0
457            do t1 = 1,25
458
459                !write this point to the first matrix
460                MAT(2,4) = this(t1,1) !x4
461                MAT(3,4) = this(t1,2) !y4
462                MAT(4,4) = this(t1,3) !z4
463
464                !compute the first determinant
465                call M44DET(MAT,det1)
466
467                !find another point on this line
468                x5(1) = this(t1,1) + Txy(1)!*0.1 !x5
469                x5(2) = this(t1,2) + Txy(2)!*0.1 !y5
470                x5(3) = this(t1,2) + Txy(3)!*0.1 !z5
471
472                !write this point to the second matrix
```

```
473                    MAT2(2,4) = x5(1) - this(t1,1) !x5-x4
474                    MAT2(3,4) = x5(2) - this(t1,2) !y5-y4
475                    MAT2(4,4) = x5(3) - this(t1,3) !z5-z4
476
477                    !compute the second determinant
478                    call M44DET(MAT2,det2)
479
480                    !find the intersection point
481                    pt = -1*(det1/det2)
482                    db(1) = this(t1,1) + (x5(1) - this(t1,1))
                          *pt
483                    db(2) = this(t1,2) + (x5(2) - this(t1,2))
                          *pt
484                    db(3) = this(t1,3) + (x5(3) - this(t1,3))
                          *pt
485  C                  db(1) = x4 - (x5 - x4)*t
486  C                  db(2) = y4 - (y5 - y4)*t
487  C                  db(3) = z4 - (z5 - z4)*t
488
489
490                    do q1 = 1,25
491
492                !determine the distance between the
                       intersection point and the point on that
                       helisotat
493                      Td(1) = db(1) - that(q1,1)
494                      Td(2) = db(2) - that(q1,2)
495                      Td(3) = db(3) - that(q1,3)
496
497                      temp1 = Td(1)**2 + Td(2)**2 + Td(3)**2
498                      temp1 = sqrt(temp1)
499
500                      dist = temp1
501
502                      if (dist.le.dcd*0.5) then
503  !                        if line(of this point t) comes close
         to (that point q)
504                        flag1 = 1
505
506                      endif
507                    end do
508                    if (flag1 .eq. 1) then
509                      r2 = r2 + 1
510                    endif
511                    flag1 = 0
512                end do
513
514                n_b = n_b*(1 - r2/25)
515
```

```fortran
516                    endif
517                 endif
518
519   c Shading --------------------
520
521              !Point to a line (distance between sun vector line
                    and heliostat k)
522              db(1) =   Sxy(2)*xa(3) - Sxy(3)*xa(2)
523              db(2) = -(Sxy(1)*xa(3) - Sxy(3)*xa(1))
524              db(3) =   Sxy(1)*xa(2) - Sxy(2)*xa(1)
525
526              temp1 = db(1)**2 + db(2)**2 + db(3)**2
527              temp1 = sqrt(temp1)
528
529              temp2 = Sxy(1)**2 + Sxy(2)**2 + Sxy(3)**2
530              temp2 = sqrt(temp2)
531
532              db(4) = temp1/temp2
533
534              if (db(4) .lt. diag) then
535              !is this heliostat in from of the other heliostat?
536              !determine the t parameter of the parametric
                    equation of the line
537              temp1 = rc(1)*Sxy(1) + rc(2)*Sxy(2) + rc(3)*Sxy
                    (3)
538              temp2 = Sxy(1)**2 + Sxy(2)**2 + Sxy(3)**2
539              t = temp1/temp2
540
541   c          write (*,*) t
542
543              if (t .gt. 0) then
544                 !Generate points on that heliostat
545                 do t1 = 1,25
546                    do q1 = 1,3
547                       that(t1,q1) = rc(q1) + dcw*pmatx(t1)*
                             thisp(q1)
548         +                         + dch*pmaty(t1)*thisu(q1)
549                    end do
550                 end do
551
552                 flag1 = 0
553                 r2 = 0
554                 do t1 = 1,25
555                    do q1 = 1,25
556                       Td(1) = this(t1,1) - that(q1,1)
557                       Td(2) = this(t1,2) - that(q1,2)
558                       Td(3) = this(t1,3) - that(q1,3)
559
560                       db(1) =   Sxy(2)*Td(3) - Sxy(3)*Td(2)
```

```
561                          db(2) =  -(Sxy(1)*Td(3) - Sxy(3)*Td(1))
562                          db(3) =   Sxy(1)*Td(2) - Sxy(2)*Td(1)
563
564                          temp1 = db(1)**2 + db(2)**2 + db(3)**2
565                          temp1 = sqrt(temp1)
566                          temp2 = Sxy(1)**2 + Sxy(2)**2 + Sxy(3)
                                 **2
567                          temp2 = sqrt(temp2)
568
569                          dist = temp1/temp2
570
571                          if (dist.le.dcd*n_c) then
572     !                       if line(of this point t) comes close
        to (that point q)
573                              flag1 = 1
574                          endif
575                        end do
576                        if (flag1 .eq. 1) then
577                          r2 = r2 + 1
578                        endif
579                        flag1 = 0
580                      end do
581
582                      n_s = n_s*(1 - r2/25)
583                    endif
584
585                endif
586
587            endif ! (k .ne. j)
588     40         continue
589
590
591         n_tot = n_c*n_at(j)*n_sp(j)*n_b*n_s
592
593         Area(i) = Area(i) + d4*d5*n_tot
594
595
596     100     continue
597             flag2 = 1 !attenuation and spillage need no longer be
                   calculated
598
599         endif
600
601     !---------------------------------
602
603         Energy = Energy - ruser(1,i)*Area(i) ! + ruser(3,i)
604         !DNI is in ruser(1,:)
605         !losses are in ruser(3,:)
606
```

```
607
608   1000  continue
609
610
611   c End of Objective Function------------
612   c9999  stop
613         return
614         end subroutine heliostat
615
616
617         subroutine dsinput (ruser,fdata)
618
619         integer fdata(*)
620         double precision Wind(192), Tamb(192),Losses(192),
621        +        Qconv, Qrad, Pr, rho, em,
622        +        Re,
623        +        pi, mu, sig, h, k, Nu, A,
624        +        ruser(3,600)
625
626   !DNI Data:
627         ruser(1,1)=0
628         ruser(1,2)=0
629         ...
630         ruser(1,191)=66
631         ruser(1,192)=0
632
633   !Sun Vector Data:
634         ruser(2,1)=0.837245769830
635         ruser(2,2)=0.915963426946
636         ...
637         ruser(2,575)=0.074520459253
638         ruser(2,576)=-0.129568108675
639
640   !Wind Data:
641         Wind(1)=2.3
642         Wind(2)=2.0
643         ...
644         Wind(191)=3.0
645         Wind(192)=2.7
646
647   !Temperature Data:
648         Tamb(1)=20.2
649         Tamb(2)=20.1
650         ...
651         Tamb(190)=32.0
652         Tamb(191)=31.1
653         Tamb(192)=30.1
654
655   !Losses
```

```
656        pi = 3.14159265359d0
657        Ts = (560+280)/2 !(Tin + Tout)/0.5
658        mu = 1.8d0!*(10**(-5)) !kg/m.s (cengal 884, 4th Ed.)
659        mu = mu/(100000)
660        Pr = 0.72 !%W/m-K (Cengal 884 - properties of air 4th ed
               .)
661        rho = 1.2 !%W/m-K (Cengal 884 - properties of air 4th ed
               .)
662        k = 0.026 !%W/m-K (Cengal 884 - properties of air)
663        em = 1 !cengal 4th pg. 711
664        sig = 5.670d0/100000000
665
666   c External cylindrical receiver------
667        A = pi*fdata(8)*fdata(8)
668
669        do 110 i = 1, fdata(7)
670   c Convection
671            Re = rho*Wind(i)*fdata(8)/mu
672            Nu = 0.027*(Re**(0.805))*(Pr**(1/3)) !Cengal 4th ed.
                   pg. 435
673            h = Nu*k/fdata(8)
674            Qconv = h*A*(Ts-Tamb(i)) !cengal 4th ed. pg.
675   c Radiation
676            Qrad = em*A*sig*((Ts+273)**4-(Tamb(i)+273)**4) !
                   cengal 4th ed. pg.711
677            ruser(3,i) = Qconv + Qrad
678   110    continue
679   c -----------------------------
680
681        return
682        end subroutine dsinput
683
684        subroutine topology(Field,Fieldz,n2)
685
686        integer n2
687        double precision x, y, z, Field(*),Fieldz(*)
688
689        do 123 i = 1, n2
690            x = Field(i)
691            y = Field(i+n2)
692   ! Default (No Slope)
693            z = 0
694   ! Slope down 1m/100m East to West
695   !          z = -x/100
696   ! Slope down 1m/100m North to South
697   !          z = -y/100
698   ! Slope down 1m/100 NE to SW
699   !          z = ?
700            Fieldz(i) = z
```

```fortran
701
702   123   continue
703
704         return
705         end subroutine topology
```

B.2   Octave Code

```octave
1    clear Z
2    clear Ys
3    clear Z_deg
4    clear Solar
5    clear E
6
7    %---Inputs--------
8    Ls = 30;
9    Ll = 21;
10   p = 28.433;  %negative
11   p = -1*p*pi/180;
12   %-----------------
13
14   %---Data Files----
15   year_Day
16   hour
17   %-----------------
18
19
20   for i = 1:8760
21      #Day angle------------
22            B = (Year_Day(i) - 1)*360/365.242;
23            B = B*pi/180;
24            #--------------------
25
26
27      #--Equation of time---
28            E(i) = 229.2*(0.000075 + 0.001868*cos(B) - 0.032077*
                  sin(B) - 0.014615*cos(2*B) - 0.04089*sin(2*B));
29            #-----------------
30
31
32      #Solar Time--------
33            Solar(i) = Hour(i) + (4*(Ll - Ls) + E(i))/60;
34            #-----------------
35
36
37            #--Hour Angle-----
38            w = (Solar(i)/24 - 0.5)*360;
39            w = w*pi/180;
40            #-----------------
```

```
41
42
43    #--Delta (declination)--
44    d = 0.006918 - 0.399912*cos(B) + 0.070257*sin(B) - 0.006758*
          cos(2*B) + 0.000907*sin(2*B) - 0.002679*cos(3*B) +
          0.00148*sin(3*B);
45    #------------------
46
47
48    #Zenith Angle------
49        z  = pi/2 - asin(cos(p)*cos(d)*cos(w) + sin(p)*sin(d))
             ;
50
51        %if z > pi/2
52        %        Z(i,1) = pi/2;
53        %else
54                Z(i,1) = z;
55        %endif
56        #------------------
57
58
59        #Azimuth Angle----
60        ys = sign(w)*abs(acos((cos(z)*sin(p) - sin(d))/(sin(z)
             *cos(p))));
61
62    %if ys<0
63        %Ys(i,1) = ys + 2*pi;
64    %else
65      Ys(i,1) = ys;
66    %endif
67
68        Ydeg(i,1) = Ys(i,1)*180/pi;
69        #------------------
70
71
72        #Sun Unit Vector--
73        alpha(i) = pi/2 - Z(i,1);
74
75        S(i,1) = cos(alpha(i))*(-sin(Ys(i,1)));   #S(i-hat) or
             Se (e for east)
76        S(i,2) = cos(alpha(i))*(-cos(Ys(i,1)));   #S(j-hat) or
             Sn (z for north)
77        S(i,3) = sin(alpha(i));                        #S(k-hat
             ) or Sz (e for zenith)
78        #------------------
79
80 end
```

APPENDIX C

OPTIMIZATION ALGORITHM

A description of the SAOi algorithm procedure is given here. The description is adapted from Groenwold and Etman [48].

Algorithm Procedure

Let k represent an outer iteration counter. Then, using either a dual subproblem or a QP subproblem, algorithm SAOi proceeds as follows. For the sake of brevity and ease, the presentation is superficial; interested readers are referred to the cited literature for details.

1. **Initialization:**
   Set $k = 0$.

2. **Simulation and sensitivity analysis:**
   Compute $f_j(\boldsymbol{x}^{\{0\}})$, $\boldsymbol{\nabla} f_j(\boldsymbol{x}^{\{0\}})$, $j = 0, 1, 2, \cdots, m$.

3. **Construct the approximations:**
   Reinitialize inner-loop specific parameters, and then construct the approximate functions $\tilde{f}_j(x)$ at $x^{\{k\}}$, $j = 0, 1, 2, \cdots m$.

4. **Approximate optimization:**
   Construct a local approximate subproblem based on equation 5.9. Solve the subproblem to arrive at $(x^{\{k*\}}, \lambda^{\{k*\}})$.

5. **Simulation analysis:**
   Compute $f_j(x^{\{k*\}})$, $j = 0, 1, 2, \cdots, m$.

6. **Test if $x^{\{k*\}}$ is acceptable:**
   If satisfied, GOTO Step 8; else CONTINUE.

7. **Effect an inner loop:**
   Adjust subproblem $[k]$ such that the likeliness of arriving at an acceptable solution $x^{\{k*\}}$ increases. GOTO Step 4.

8. **Move to the new iterate:**
   Set $x^{\{k+1\}} := x^{\{k*\}}$.

9. **Test for convergence:**
   If satisfied, STOP; else CONTINUE.

10. **Simulation sensitivity analysis:**
    Compute $\nabla f_j(x^{\{k+1\}})$, $j = 0, 1, 2, \cdots, m$.

11. **Initiate a new outer loop:**
    Set $k := k + 1$ and GOTO Step 3.

It is more precise to use the notation $x^{\{k,l\}}$, with l an inner iteration counter, rather than $x^{\{k\}}$. The latter however is retained for the sake of brevity, and the meaning, at least, is clear.

Step 6 provides the mechanism for global convergence. That is, the solution $x^{\{k*\}}$ to subproblem $[k]$ is only accepted to become the new iterate $x^{\{k+1\}}$ if sufficient improvement is realized. Typically this improvement is expressed in terms of a merit function, or in terms of a filter Pareto front, balancing the contribution of objective function and constraint violation. If the candidate iterate happens to be unacceptable, subproblem $[k]$ is adjusted and re-solved. The adjustment of the subproblem should be such that it becomes more likely that the solution to the adjusted subproblem will pass the acceptability test. A well-known approach to effect this, is to include trust region constraints in the subproblem, and to subsequently reduce the trust region in case of failure of the acceptability test.

BIBLIOGRAPHY

[1] Torresol Energy: Gemasolar thermosolar plant [Online]. Available at: `http://tinyurl.com/3utoytt`, [11 November 2013].

[2] Scheer, H.: *Energy Autonomy: The Economic, Social and Technological Case for Renewable Energy.* Earthscan, 2006.

[3] Visagie, E. and Prasad, G.: Renewable energy technologies for poverty alleviation—South Africa: biodiesel and solar water heaters. Energy Research Centre, University of Cape Town, South Africa, 2006.

[4] Scheer, H.: *A Solar Manifesto.* James and James Ltd., London, 2001.

[5] Stine, W.B. and Geyer, M.: Power from the Sun [Online] (2001). Available at: `www.powerfromthesun.net`, [11 November 2013].

[6] Pitz-Paal, R.: Concentrating solar power. *Energy: Improved, Sustainable and Clean Options for our Planet,* pp. 171–192, 2008.

[7] United States Department of Labor: Careers in Solar Power [Online]. Available at: `http://www.bls.gov/green/solar_power`, [11 November 2013].

[8] Kuravi, S., Trahan, J., Goswami, D.Y., Rahman, M.M. and Stefanakos, E.K.: Thermal energy storage technologies and systems for concentrating solar power

plants. *Progress in Energy and Combustion Science*, vol. 39, pp. 286–319, 2013.

[9] Murphy, L.M. and May, E.K.: Steam generation in line-focus solar collectors: a comparative assessment of thermal performance, operating stability, and cost issues. Tech. Rep., Solar Energy Research Inst., Golden, CO (USA), 1982.

[10] Mehos, M.: Concentrating solar power. In: Proceedings of the 2008 AIP Conference, 2008.

[11] Schell, S.: Design and evaluation of eSolar's heliostat fields. *Solar Energy*, vol. 85, no. 4, pp. 614–619, April 2011.

[12] Yogev, A., Kribus, A., Epstein, M. and Kogan, A.: Solar "tower reflector" systems: a new approach for high-temperature solar plants. *International journal of hydrogen energy*, vol. 23, no. 4, pp. 239–245, 1998.

[13] Danielli, A., Yatir, Y. and Mor, O.: Improving the optical efficiency of a concentrated solar power field using a concatenated micro-tower configuration. *Solar Energy*, vol. 85, no. 5, pp. 931–937, May 2011.

[14] Kolb, G.J., Ho, C.K., Mancini, T.R. and Gary, J.A.: Power Tower Technology Roadmap and Cost Reduction Plan. Tech. Rep., Sandia National Laboratories, 2011.

[15] Garcia, P., Ferriere, A. and Bezian, J.-J.: Codes for solar flux calculation dedicated to central receiver system applications: A comparative review. *Solar Energy*, vol. 82, no. 3, pp. 189–197, March 2008.

[16] Shuai, Y., Xia, X.-L. and Tan, H.-P.: Radiation performance of dish solar concentrator/cavity receiver systems. *Solar Energy*, vol. 82, pp. 13–21, 2008.

[17] Bode, S.-J. and Gauché, P.: Review of optical software for use in concentrating solar power systems. In: Proceedings Southern African Solar Energy Conference 2012, Stellenbosch, South Africa, 2012.

[18] Blanco, M.: Current status of Tonatiuh—A computer program for the simulation of solar concentrating systems. In: Proceedings 16th Workshop on Crystalline Silicon Solar Cells & Modules, Denver, CO, 2006.

[19] Leonardi, E. and Aguanno, B.D.: CRS4-2 : A numerical code for the calculation of the solar power collected in a central receiver system. *Energy*, vol. 36, no. 8, pp. 4828–4837, 2011. ISSN 0360-5442.

[20] Noone, C.J., Torrilhon, M. and Mitsos, A.: Heliostat field optimization: A new computationally efficient model and biomimetic layout. *Solar Energy*, vol. 86, no. 2, pp. 792–803, February 2012.

[21] Collado, F.J.: Preliminary design of surrounding heliostat fields. *Renewable Energy*, vol. 34, no. 5, pp. 1359 – 1363, 2009.

[22] Collado, F.J.: Quick evaluation of the annual heliostat field efficiency. *Solar Energy*, vol. 82, no. 4, pp. 379–384, April 2008.

[23] Gauché, P., von Backström, T.W. and Brent, A.C.: CSP Modeling methodology for macro decision making—emphasis on the central receiver type. In: Proceedings SolarPACES 2011, 20–23 Sept. 2011. Granada, Spain, 2011.

[24] Erol, O.K. and Eksin, I.: A new optimization method: big bang–big crunch. *Advances in Engineering Software*, vol. 37, no. 2, pp. 106–111, 2006.

[25] Venter, G.: Review of optimization techniques. *Encyclopedia of aerospace engineering*, 2010.

[26] Bischof, C., Khademi, P., Mauer, A. and Carle, A.: Adifor 2.0: Automatic differentiation of fortran 77 programs. *Computational Science & Engineering, IEEE*, vol. 3, no. 3, pp. 18–32, 1996.

[27] Rao, S.S.: *Engineering Optimization: Theory and Practice*. Wiley & Sons, Hoboken, New Jersey, 2009.

[28] Reyes-Sierra, M. and Coello, C.C.: Multi-objective particle swarm optimizers: A survey of the state-of-the-art. *International journal of computational intelligence research*, vol. 2, no. 3, pp. 287–308, 2006.

[29] Nakayama, H., Yun, Y. and Yoon, M.: *Sequential approximate multiobjective optimization using computational intelligence*. Springer, 2009.

[30] Groenwold, A.A., Etman, L.F.P. and Wood, D.: Approximate approximations for SAO. *Structural and Multidisciplinary Optimization*, , no. 41, pp. 39–56, 2010.

[31] Nakayama, H., Yun, Y. and Yoon, M.: Sequential approximate multiobjective optimization using computational intelligence. In: *9th Int. Symp. Oper. Res. Appl., Chengdu, China*. 2010.

[32] Barthelemy, J.-F. and Haftka, R.T.: Approximation concepts for optimum structural designa review. *Structural optimization*, vol. 5, no. 3, pp. 129–144, 1993.

[33] Sánchez, M. and Romero, M.: Methodology for generation of heliostat field layout in central receiver systems based on yearly normalized energy surfaces. *Solar Energy*, vol. 80, no. 7, pp. 861–874, July 2006.

[34] Buck, R.: Heliostat field layout using non-restricted optimization. In: Proceedings SolarPACES 2012, September 11-24, Marrakech, Morocco, 2012.

[35] Kistler, B.: A user's manual for DELSOL3. Tech. Rep., Sandia National Labs., Livermore, CA (USA), 1986.

[36] Pitz-Paal, R., Botero, N. and Steinfeld, A.: Heliostat field layout optimization for high-temperature solar thermochemical processing. *Solar Energy*, vol. 85, pp. 334–343, 2011.

[37] Green Wombat: Transmission constraints derail California solar project [Online]. Available at: `http://thegreenwombat.com/category/esolar/`, [11 November 2013].

[38] Landman, W. and Gauché, P.: Sensitivity Analysis of a Curved Heliostat Facet Profile. In: Proceedings SolarPACES 2012, Marrakech, Morocco, 2012.

[39] Eaton, J.W., Bateman, D. and Hauberg, S.: *GNU Octave*. Free Software Foundation, 1997.

[40] Metcalf, M., Reid, J.K. and Cohen, M.: *Fortran 95/2003 Explained*, vol. 416. Oxford University Press Oxford, 2004.

[41] Biran, D.A. and Biran, A.: *MATLAB 5 for Engineers*. Addison-Wesley Longman Publishing Co., Inc., 1999.

[42] Duffie, J.A. and Beckman, W.A.: *Solar Engineering of Thermal Processes*. 3rd edn. John Wiley & Sons, Inc., New Jersey, 2006.

[43] Blanco, M., Mutuberria, A. and Martinez, D.: Experimental validation of Tonatiuh using the Plataforma Solar de Almeria secondary concentrator test campaign. In: Proceedings 2010 International SolarPACES Symposium, 21-24 September, Perpignan, France, 2010.

[44] Stewart, J.: *Calculus: Early Transcendentals*. Thomson Books, 2010.

[45] Weisstein, E.W.: MathWorld—A Wolfram Resource. *Line-Plane Intersection* [Online]. Available at: `http://mathworld.wolfram.com/Line-PlaneIntersection.html`, [11 November 2013].

[46] Sargent & Lundy LLC Consulting Group Chicago, Illinois: Assessment of parabolic trough and power tower solar technology cost and performance forecasts. Tech. Rep., National Renewable Energy Laboratory (NREL), Chicago, Illinois, October 2003.

[47] Wendelin, T.: Soltrace: a new optical modeling tool for concentrating solar optics. In: *ASME 2003 International Solar Energy Conference*, pp. 253–260. American Society of Mechanical Engineers, 2003.

[48] Groenwold, A.A. and Etman, L.F.P.: The 'Not-So-Short' manual for the SAOi algorithm, Version 0.8.10. Tech. Rep., University of Stellenbosch, Department of Mechanical and Mechatronic Engineering, Stellenbosch, South Africa, 13 October 2012.

[49] Thermosol Glass: Concentrated Solar Power—Power Tower [Online]. Available at: `http://tinyurl.com/lso2odk`, [11 November 2013].

[50] Etman, L.F.P., Groenwold, A.A. and Rooda, J.: On diagonal QP subproblems for sequential approximate optimization. In: *Eighth World Congress on Structural and Multidisciplinary Optimization, paper 1065*. Lisboa, Portugal, 2009.

[51] Groenwold, A.A., Etman, L.F.P. and Rooda, J.: First-order sequential convex programming using approximate diagonal QP subproblems. *Structural and Multidisciplinary Optimization*, , no. 45, pp. 479–488, 2012.

[52] Duysinx, P., Zhang, W., Fleury, C., Nguyen, V. and Haubruge, S.: A new separable approximation scheme for topological problems and optimization problems characterized by a large number of design variables. In: Ollhoff, N. and Rozvany, G. (eds.), *First World Congress on Structural Multidisciplinary Optimization*, pp. 1–8. Goslar, Germany, 1995.

[53] NREL: Concentrating Solar Power Projects—PS10 [Online]. Available at: `http://tinyurl.com/kfpelqb`, [11 November 2013].

[54] Greenpeace: Usina solar térmica PS10 [Online]. Available at: `http://tinyurl.com/lwzqyx2`, [11 November 2013].

[55] Fernández, D.V.: PS10: a 11.0-MWe Solar Tower Power Plant with Saturated Steam Receiver. Tech. Rep., Solucar, 2004.

www.ingramcontent.com/pod-product-compliance
Lightning Source LLC
Chambersburg PA
CBHW041310210326
41599CB00003B/49